Failsafe Control Systems

Failsafe Control Systems

Applications and emergency management

UNICOM

APPLIED INFORMATION TECHNOLOGY REPORTS

Edited by **Kevin Warwick**
Professor of Cybernetics, University of Reading

and

Ming T. Tham
Department of Chemical and Process Engineering,
University of Newcastle upon Tyne

CHAPMAN AND HALL
LONDON • NEW YORK • TOKYO • MELBOURNE • MADRAS

UK	Chapman and Hall, 2–6 Boundary Row, London SE1 8HN
USA	Van Nostrand Reinhold, 115 5th Avenue, New York NY10003
JAPAN	Chapman and Hall Japan, Thomson Publishing Japan, Hirakawacho Nemoto Building, 7F, 1-7-11 Hirakawa-cho, Chiyoda-ku, Tokyo 102
AUSTRALIA	Chapman and Hall Australia, Thomas Nelson Australia, 480 La Trobe Street, PO Box 4725, Melbourne 3000
INDIA	Chapman and Hall India, R. Seshadri, 32 Second Main Road, CIT East, Madras 600 035

First edition 1991

Softcover reprint of the hardcover 1st edition 1991

T.J. Press (Padstow) Ltd, Padstow, Cornwall

ISBN-13: 978-94-010-6677-8 e-ISBN-13: 978-94-009-0429-3
DOI: 10.1007/978-94-009-0429-3

British Library Cataloguing in Publication Data

Failsafe control systems..
1. Control systems
I. Warwick, K. (Kevin) II. Tham, Ming T.
629.8

Library of Congress Cataloging-in-Publication Data

Failsafe control systems: applications and emergency management / edited by Kevin Warwick and Ming T. Tham. — 1st ed.
p. cm.
Includes bibliographical references and index.

1. Automatic control—Reliability. I. Warwick, K. II. Tham, Ming T.
TJ213.95.F35 1990
629.8—dc20

Contents

Contributors

Dr P. Andow
KBC Process Automation
Chilworth Research Centre
Southampton

P. Ashmole
High Voltage Technology Section
Research and Development Centre
Kelvin Avenue
Leatherhead
Surrey
KT22 7ST

R. Bell
Health and Safety Executive
Magdalen House
Stanley Precinct
Bootle
Merseyside
L20 3QZ

Dr R. Clarke
Nuclear Electric
Bridgewater Road
Bedminster Down
Bristol
BS13 8AN

Dr D. Embrey
Human Reliability Associates Ltd
1 School House
Higher Lane
Dalton
Wigan
WN8 7RP

Professor M. Grauer
Department of Computer Science
University of Dortmund
Post Box 500500
D-4600 Dortmund
West Germany

P. A. L. Ham
Heaton Works
NEI Parsons Ltd
Newcastle Upon Tyne
NE6 2YL

Dr D. J. Holding
Department of Engineering
and Applied Physics
University of Aston
Aston Triangle
Birmingham
B4 7ET

D. Jackson
National Engineering Laboratory
Department of Trade and Industry
East Kilbride
Glasgow
G75 0QU

I. Jardine
Baker Jardine and Associates Ltd
Lakeview
Cally Drive
Gatehouse-of-Fleet
Castle Douglas
DG7 2DJ

M. Kangethe
Department of Electronics
University of York
Heslington
York
YO1 5DD

G. C. Meggitt
Safety and Reliability Directorate
UKAEA
Wigshaw Lane
Culcheth
Warrington
WA3 4NE

R. W. Milne
Intelligent Applications Ltd
Kirton Business Centre
Kirk Lane
Livingstone Village
West Lothian
EH54 7AY

M. F. Pantony
Health and Safety Executive
Magdalen House
Stanley Precinct
Bootle
Merseyside
L20 3QZ

Dr R. Patton
Department of Electronics
University of York
Heslington
York
YO1 5DD

Professor M. Rijckaert
Instituut Voor Chemie
Ingenieurstechn
De Croylaan 2
3030 Heverlee
Belgium

Dr D. J. Sandoz
VUMAN Ltd
Enterprise House
Manchester Science Park
Lloyd Street North
Manchester
M15 4EN

D. J. Smith
Technis
26 Orchard Drive
Tonbridge
Kent
TN10 4LG

Dr M. T. Tham
Department of Chemical Engineering
University of Newcastle upon Tyne
Newcastle upon Tyne
NE1 7RU

Professor K. Warwick
Department of Cybernetics
University of Reading
Whiteknights
Reading RG6 2AL

M. de Winter
Instituut Voor Chemie
Ingenieurstechn
De Croylaan 2
3030 Heverlee
Belgium

T. G. Wolfmaier
Human Reliability Associates Ltd
1 School House
Higher Lane
Dalton
Wigan
WN8 7RP

ABOUT THE EDITORS

KEVIN WARWICK is Professor of Cybernetics and Head of the Department of Cybernetics, University of Reading. Previously he held appointments at Warwick University, Oxford University, Imperial College London and Newcastle University as well as being employed by British Telecom for six years. Profesor Warwick is a Chartered Engineer and is Honorary Editor of the IEE Proceedings, Part D, on Control Theory and Applications. He authored the book *Control Systems: An Introduction*, and has edited several books including the successful *Industrial Digital Control Systems*. He has published over ninety papers in the areas of his research interests which include intelligent control systems, sensor integration and intelligent manufacturing.

MING T. THAM is a lecturer in the Department of Chemical and Process Engineering at the University of Newcastle upon Tyne. He graduated with a first class BSc (Hons) degree in Chemical Engineering and followed this up with a PhD in adaptive control systems from Newcastle University. His research interest is in the development and application of advanced control methodologies to the process industries.

PREFACE

This book is intended to provide an introduction to and an overview of the techniques and methodologies involved in the design and application of fail-safe control systems and emergency management. The book is aimed at practising engineers, project manager and engineering students who have a basic knowledge of computers, control systems and some mathematics and who wish to become informed about the various aspects of fail-safe control. What follows is not concerned with the proffering of highly abstract ideas based on theoretical assumptions, but rather is firmly centred on industrial problems and industrial practice. Indeed many of the chapters are founded on the description of industrial procedures employed to deal with actual problems areas.

The book is based on a successful seminar on Fail-Safe Control Systems held in June 1988 at the National Liberal Club, Whitehall Place, London. The seminar, organized by Unicom Seminars, was well supported by both industrialists and academics alike and it was due to the positive response of the participants, with regard to the material presented, that this book has been put together. The text provides an overall balanced perspective on the theme of fail-safe control and may be used as a series of reference articles for industrial uptake or as support material on study courses in which safety in control forms an important aspect.

The book commences with a brief introduction of the subject area, in terms of both theory and practice, and this is followed, in Chapters 2, 3 and 4 by a look at software fault tolerance. This includes in Chapter 4 however, a consideration of operator support systems for fault diagnosis and information displays.

Chapters 5 and 8 are concerned with various aspects of fault tolerant control systems and Chapters 9 to 11 discuss fault diagnosis techniques, Chapter 11 in fact describing the prediction of failures by the use of reliability analyses and performance simulation.

Chapters 12 to 13 look at emergency management, while Chapters 14 and 15 give an insight into the use of expert systems in safety critical systems. This in fact leads on to the more general field of intelligent control systems which is taken a stage further in Chapter 16 with a consideration of intelligent process control. Finally Chapter 17 investigates how advanced control systems can be safely incorporated to meet desired objectives in terms of quality control and security.

In conclusion the Editors would like to thank all the authors for their contributions, and their promptness in forwarding the chapters for presentation. The Editors would like to give a special mention to Professor Manfred Grauer and Marylene De Winter who travelled some distance to give their presentations at the London seminar. Also on behalf of all the authors, the Editors would like to thank those at Unicom Seminars who were responsible for the production of the seminar and latterly this book, namely Gautam Mitra and Boris Sedacca, but particularly Julie Valentine.

Kevin Warwick
Ming T. Tham
January 1989

ABOUT THIS BOOK

This book provides an overview and update of techniques used in emergency management and fail-safe control system design. As well as giving tutorial type coverage of the field, describing the general concepts and basic ideas employed, many of the chapters include details of applications and discuss the problems from a pragmatic viewpoint based on practical experience.

The design of fault tolerant control systems is considered at length and safety related aspects are highlighted. The application of fault tolerant procedures is presented by means of several real examples and links are drawn between theoretically sound techniques and their resultant performance when operated on-line. The related area of fault diagnosis is also discussed in several chapters; here, emphasis is placed on the use of diagnostics, as predictive mechanisms for avoiding future problem areas.

Chapter 1

FAIL-SAFE CONTROL SYSTEMS: AN INTRODUCTION

M.T. Tham and K. Warwick* +
**University of Newcastle upon Tyne*
+ *University of Reading*

Techniques and procedures employed in control applications have changed significantly over the last decade, one reason being the wide-spread automation of monitoring and control operations by means of digital computers. Indeed, computer-controlled systems, providing a spectrum of flexible and powerful software tools are now commonplace. However, it is often the case that conventional controllers have merely been replaced or updated by a computer-based scheme. Such a philosophy ignores the advantages and improvements that can be obtained in overall performance and plant operation.

Nevertheless, advances in microelectronic technology have increased the power and flexibility of microprocessor based instrumentation and hence opened new possibilities. Familiar stand-alone analogue single-input single-output control loop configurations are rapidly becoming obsolete. They are being replaced by electronic controllers with multi-input capabilities, devices with advanced control capabilities, and dedicated process supervision units all of which may be linked to a central computing resource. Typically front-end control devices, alarms and associated management systems, data acquisition units and so on, are linked to supervisory units on the next operating level. These may in turn be accessed by higher levels for data acquisition, archiving operations and management analyses.

The trend is towards highly integrated and sophisticated computer systems for process management, supervision and operation. At the same time, controller requirements have become more stringent, both in terms of safety and plant productivity. In plants processing hazardous products, the control system must be able to initiate shut-down procedures upon detection of process failures. Device failures may also occur. Under such circumstances, a further requirement is for the device to be able to operate in the normal manner until shut-down is completed. Plant shut-downs, however, are expensive affairs. In less critical situations, economic considerations dictate that the process control system must present a high degree of availability. In other words, the system should be able to maintain normal operation until the device fault is rectified.

A foremost consideration in the development and design of process control systems, therefore, is the ability of the sub-units as well as that of the ensemble to tolerate system faults. This requirement is not restricted to the hardware, but also extends to the software environment. System reliability is made more critical in highly automated and integrated settings, where it is possible that faults may propagate undetected and may be amplified with disastrous consequences.

The requirement for reliable or fault tolerant systems is not new. Traditionally, system reliability has been maintained by the specification of hot standby units in duplex or triplex configurations managed by voting functions. For example, several controllers may be operated in parallel to monitor and control the same process variable. Similarly, multiple sensors may be used to measure one process variable. The voting function then interrogates each multiplexed configuration to establish which controller is providing the correct action, or which sensor is providing the most reliable measurement. This technique requires additional hardware, all of which must be linked together, resulting in costly and complex control systems.

The quest for cost effective reliable process control systems, and systems in general, has generated a new field of study: reliability engineering. The methodology underlying the current approach is that of analytical or functional redundancy, involving reliability modelling and analysis. The former enables the prediction of future device failures. Reliability prediction requires the determination of failure mechanisms followed by quantification of failure rates, viz. reliability analysis. These results can then be used for timetabling maintenance schedules, design of fault tolerant systems and for specifying the degree of redundancy of components. This is contrasted to providing reliability via redundancy of devices – a more expensive option.

Apart from device failure prediction prior to design, fault diagnosis techniques may also be used on-line to detect and localize sensor and component faults. Such techniques are usually applied to the detection of faults occurring with the process being controlled or monitored, and associated ancillary equipment such as valves, pumps, compressors and so on. Popular approaches include the use of Kalman filtering techniques, Luenberger observers and even "black-box" models based on input-output data. These algorithms provide information about the process from which process states may be extracted for fault detection and classification.

Detection that a fault has occurred or is likely to occur in a plant, tracing the cause of the problem, and instigating appropriate action, are all features which can be readily achieved by means of active computer-resident knowledge bases. In practice, the majority of knowledge-based systems require the operator to enter information into the base once this has been either read directly, or calculated from measured values. The use of on-line data collection hardware, in this respect, is now becoming more widespread.

Allied to the requirements of safe plant operation, is the installation of alarm and trip systems which are part and parcel of all process control environments. A feature of their design is that critical parts of the process are allocated higher priorities in terms of alarm management tasks. However, as process plants become more integrated, such systems become correspondingly more complex. Thus, a further consideration is a good operator interface: the system must present information in a manner that is easily understood and operated on by plant personnel. An alternative automated approach, is the use of "expert systems" for management of alarms and emergency tasks. This technique has attracted substantial interest, and much development work is being carried out.

Clearly, there are many considerations which have to be taken into account in the design of fail-safe control systems. Firstly, many fault tolerant techniques are software driven, and here reliability problems can also be encountered. The integrity of process control systems can be undermined through a poor operator/machine interface, and thus the reliability of software procedures and their interaction with human operators must be given due attention. Next, the fault tolerant control systems design must be performed bearing in mind process constraints and operational safety aspects. The related area of fault diagnosis is of equal significance where the emphasis is on the use of predictive mechanisms in order to avoid future problems and to initiate preventive maintenance. Emergency management is just as important and here, the potential afforded by artificial intelligence techniques should be explored, to provide an expert systems functional task operative. Finally, the use of advanced, intelligent control systems is gradually becoming more acceptable as it is realized that trust can be placed in adaptive control forms, even for the most troublesome and complex of plants, on condition that the controller is adequately configured and that an appropriate human expert is consulted.

REFERENCES

1 Warwick, K. and Tham, M.T., "Emergencies and Fail-Safe Control", Control and Instrumentation, Vol. 20, No. 6, pp. 67-70, 1988.

2 Warwick, K. and Rees, D. (Eds), "Industrial Digital Control Systems", Comprehensively Revised 2nd edition, Peter Peregrinus Ltd., 1988.

3 Warwick, K., "Control Systems: An Introduction", Prentice-Hall Inc., 1988.

Chapter 2

SOFTWARE FAULT TOLERANCE

David J. Holding
Aston University , Birmingham, UK

2.1 INTRODUCTION

Programmable electronic systems such as microprocessors or transputers offer high computational power, high reliability and low power consumption at a low cost. Their use in industry has increased significantly in recent years, particularly in embedded applications such as real-time instrumentation and control systems.

It is important that such systems are able to perform in a reliable and safe manner over considerable periods of time. This can only be achieved if each system is designed correctly, implemented properly, and has the robustness required to survive in a particular operating environment. The reliability of a system can be increased by designing the system hardware and software to be fault tolerant. This chapter discusses the use of software fault tolerance in the design of high reliability systems.

2.2 WHAT IS SOFTWARE FAULT TOLERANCE?

The starting point for the design of a software based system is the derivation of the system requirement specification which states what a system should do. The designer will translate this specification into a design and will implement the design on a suitable processing architecture. An objective at all stages of the design process is to avoid introducing faults into the design. Thus, in modern software engineering, the designer is encouraged to use formal methods in the derivation of the specification and the translation of the specification into a design. Ideally, the designer should prove the correctness of the design and the design should be translated into an implementation using proven techniques and tools. Finally, the implementation should be verified to show that it is fit for its intended use.

Unfortunately, the use of formal methods requires high levels of skill and the current techniques are lengthy and are not efficient for complex systems. In practice, these techniques are not always used in embedded applications and, although all designers aim to avoid making design errors, faults are introduced into designs either explicitly, as part of a particular component, or implicitly, through the omission of a particular feature or attribute.

Errors in the specification or design of a software-based system will lead to software faults ("bugs") which will lie hidden within the system. The fault will only become apparent if a particular instantiation of the system state activates the fault and generates an error; the error may lead to the failure of the system. The number of faults in a piece of software does not increase with age, nor does the nature of a software fault change with age if the operational environment is invariant. Thus, the process of ageing of software is different from the process of ageing of hardware. This is shown by the statistics of the occurrence of errors in software [1]; the rate of occurrence of errors is highest during the development or early life of the software and, if the faults are corrected, the rate of occurrence of software errors decreases throughout the life of the software.

In an attempt to increase the reliability of the design, a designer may consider using software fault tolerance in the design of a system. Software fault tolerance techniques aim either to mask the effect of a fault or to limit the scope of a fault [2]. In the fault masking technique, an error arising from a fault is detected by comparison with correct data computed using variant paths, the error is then suppressed [3]. Alternatively, the scope of a fault may be limited by detecting an error arising from a fault and invoking an appropriate error recovery mechanism [4]. In both cases, the aim is to design a system which will meet its specification in the presence of certain classes of fault.

Software fault tolerance can be built into a system during the design process. The form of the fault tolerant structure will depend on the nature of the application, the characteristics of the software, and the architecture of the processing system. The criteria which determine when and how to use software fault tolerance are described in the following sections. In general, software fault tolerance techniques are applied economically, perhaps to protect a specific function of the system.

2.3 SEQUENTIAL, CONCURRENT AND REAL-TIME SYSTEMS

Traditionally, real-time embedded systems have been designed around a centralized computing resource such as a microprocessor. In a centralized system, the processor has immediate access to all the data in the system and has to carry out all the tasks associated with a particular application. In the simplest cases only, a simple sequential process will control the input or acquisition of data, the processing of data according to a prescribed algorithm, and the output of results. Methods for introducing fault tolerance in sequential systems are described in Section 2.6.

Generally, a processor may be required to run several tasks at the same time, as in a multi-loop control system. In these systems, the software consists of a set of processes which are active simultaneously and share the computing resource. In the traditional approach to multi-programming, each task is programmed in a high level sequential programming language and is executed under a multi-task operating system. Such multi-tasking systems are characterized by complex systems software and limited performance, particularly if the number of tasks is large. Some of these problems can be overcome by sharing or distributing the computational load between a number of processors.

In many multi-tasking applications, the concurrent tasks or processes must co-operate, through inter-process communications, to provide the overall system function. The design of fault tolerant concurrent software is much more complex because an error can migrate between processes through inter-process actions which cannot be revoked. The arbitary placement of fault tolerant structures in such systems is unlikely to result in increased reliability and may result in the progressive collapse of the whole system (the domino effect) [4]. The design of fault tolerant concurrent software is described in Section 2.7.

Real-time systems are used in many engineering applications. These systems are required to maintain synchronism with an asynchronous external system, or to respond to stimuli from such a system within a finite and specifiable delay. For example, in real-time control, a precise time-window is specified during which sensors must be sampled, a satisfactory control response computed and values output to actuators. To ensure that the system complies with the timing requirements, the performance of critical processes can be monitored using watchdog mechanisms which warn of potential violations should it appear that the process will overrun and take appropriate action to produce a timely result [5]. The design of watchdog fault tolerance mechanisms for sequential and concurrent systems is described in Sections 2.6 and 2.7 respectively.

The processors in a multi-processor system may be tightly coupled and share common memory or may be loosely coupled and interconnected by an appropriate communications network, as in a network of microprocessors or transputers. A variety of concurrent programming languages are available, each language contains variants of standard sequential and concurrent constructs and is aimed at a particular processor model [6]. For example, Modula is aimed primarily at tightly coupled processors, Occam [7] is aimed primarily at networks of loosely coupled processors, and Ada can embrace both tightly coupled and loosely coupled systems. The following sections describe software fault tolerant techniques for both centralized and distributed systems.

A distributed system is said to be truly decentralized if no single process has a complete record of the overall system state. The problem of maintaining the integrity of data in decentralized systems, and of computing distributed transactions, has been studied extensively [8, 9]. The problem of taking distributed, multi-party decisions in decentralized systems is discussed, and appropriate fault tolerant distributed decision mechanisms are described, in Section 2.8.

2.4 ANALYSIS AND SYNTHESIS TOOL

Fault tolerant structures can be used to mask or limit the extent of errors. In particular, fault tolerant techniques can cope sucessfully with many types of transient fault; they can also cope with some types of permanent fault. However, the use of fault tolerant techniques in a system should not be used as an excuse to lower the standard of design.

The design and verification of sequential, concurrent and real-time programs is not trivial. The use of formal notations and programming languages is recommended because they have an underlying mathematical structure which can be used in the formal specification of the system and in the synthesis and verification of the design, including any fault tolerant structures.

For example, state models can be generated for the six primitive constructs which are to be found in all sequential programming languages: sequence, input, output, assignment, selection (decision), and repetition (iteration). State models can be generated also for the additional constructs which are required in concurrent programming languages to initiate a number of processes in parallel, to allow inter-process communication via shared variables or message passing, to terminate asynchronous races such as those which occur when a number of processes compete for access to a non-pre-emptive resource, and to enforce mutual synchronization between two processes. In general, these requirements can be satisfied by the use of two additional constructs only [10].

Concurrent programming languages such as Modula, Concurrent Pascal, Ada and Occam provide a variety of constructs for describing concurrency. Languages which enforce synchronization by communication, and prohibit asynchronous communication, also enforce a strict discipline on the designer because errors in the synchronization logic will lead to deadlock [10]. Deadlock occurs when an active process is waiting for an event which will not occur, typically this arises from poor design. For example, the process which loses an asynchronous race for a resource will be deadlocked unless its communication for the resource is satisfied. Languages which use the more flexible, asynchronous form of communication would appear to offer a design advantage [6]. However, the additional effort needed to design a synchronous system is often more than compensated by the reduction in the effort needed to test the system because such systems are more amenable to analysis.

Considerable work has been done on the state-modelling of sequential and concurrent software using Petri nets [11, 12]. These models can be used to simulate process execution, this allows the designer to obtain important information about the dynamic behaviour and performance of the software. Analysis of the Petri net is normally carried out using reachability sets. In particular, the designer can resolve many state reachability problems and determine whether the net is prone to deadlock and requires redesign. To keep the analysis manageable it is often necessary to attempt to bound the net by using specific programming techniques and to reduce the complexity of the net by applying net-reduction techniques, such as partitioning. However, the analysis is still subject to the problem of combinatorial explosion.

2.5 AN OVERVIEW OF SOFTWARE FAULT TOLERANCE TECHNIQUES

Fault tolerance involves the use of redundancy to give the system resilience [2, 3, 13, 14, 15]. It can be used to limit the scope of the faults or mask errors so that they do not lead to failure, to protect inter-process communications in distributed systems, and to ensure the integrity of decision processes in decentralized systems. In addition, in time-critical systems, it can be used to ensure that results are computed in a timely manner.

(a) Fault masking

The traditional approach to fault tolerance is fault masking, in which the massive redundancy of software is used to mask errors. The most common technique is N-modular redundancy proposed by Avienzis [3] which involves the n-fold replication of processes. The replicated processes are operated either in serial or in parallel and the results computed by the variant paths are compared. If an error is generated by any one path, it will be detected and the erroneous result can be suppressed, thus masking the fault. The usual replication factor is three and diverse techniques are used in the programming of the triplicated paths.

N-version programming and recovery block techniques have been used successfully to provide software fault tolerance for sequential systems [4, 16].

(b) Error detection and recovery

Error detection and recovery techniques provide an alternative approach to software fault tolerance. The aim of these methods is to detect an error, to assess the damage caused by the error, and to initiate an error recovery mechanism in order to compute an error-free result with the minimum delay. The method is characterized by the economical use of redundant software.

Two recovery techniques are commonly used. The forward error recovery method [17, 18] aims to minimize delay by transforming the erroneous state directly into a correct or acceptable state. For example, a previously computed acceptable value, or a default value, can be used to replace an erroneous value in the erroneous state. In the alternative method, backward error recovery, the system backtracks through previous states until it can restore a previously saved correct state and restart processing in the forward direction. Following the restart, a diverse forward path is often used to save time or to avoid a common mode error in the forward path.

(c) Watchdog mechanisms

It is conventional to monitor the performance of a time-critical application process using a "watchdog" timer [5]. Traditionally, this is done using a real-time time-lapse counter which is preset to trip after a pre-determined time. The pre-determined trip period should be set to somewhat less than the time-critical time so that fault recovery can take place and the system can still provide a timely and satisfactory response. The timer is initiated to run concurrently with the time-critical process and the first process to complete causes the other to abort. If the application process has been designed properly, it will produce results well before the maximum allowed time and the watchdog timer will be aborted. Alternatively, a watchdog timer trip indicates the presence of a fault (which may be a software design fault or a transient or permanent malfunction of the system) and appropriate fault recovery activities should be invoked. In concurrent systems, deadlock may arise if a crude attempt is made to abort a process, and appropriate non-deadlocking process abortion mechanisms must be used as described in Section 2.7(c).

(d) The protection of communications

In a distributed system, the communications medium may not be immune to faults and must be regarded as a potential source of errors. Typical errors are erroneous messages or messages which fail to arrive. In loosely coupled systems such errors may have a serious effect because inter-process messages are used to ensure the proper synchronization and operation of the distributed systems. In synchronous systems, prompt recognition of a communication fault is essential if recovery actions are to be initiated and deadlock avoided. In real-time systems, these actions must be carried out within a critical time period. Software fault tolerance techniques can be used in concurrent real-time systems to protect the inter-process communications.

Communication transactions can be protected using watchdog timers only if there is a logical pairing of the protective mechanisms on each side of the transaction [19]. For example, if a timeout is used on the receive primitive only, the sending process will hang if the timeout operates, since the sending process has no means of determining whether the communication medium has failed. In many cases, logically paired timeout mechanisms cannot be designed without extensions to conventional concurrent programming languages [20]. This problem is generic to attempts to protect communications at the primitive or transaction level.

A solution can be obtained by decreasing the level of granularity at which protection is applied and enclosing a communication which must be protected within a conversation. One property of the conversation is the absence of communications through the sidewalls. Therefore, it is possible to protect the conversation using a watchdog timer which monitors the performance of the complete set of processes and communication transactions within a conversation. In the event of a participating process becoming unresponsive, or a failure in the communications between processes, the conversation will not terminate. The default process or conversation can then be invoked by the timer and the failure need only become perceptible after other time-critical processes succeed in producing results [21].

(e) Data integrity and transaction processing in decentralized systems

Hierarchical distributed control systems make extensive use of higher-level inter-process transactions to promulgate commands or alarms, to elicit status information, and to implement multi-party decisions [19]. In such systems, the process of making decisions is usually a centralized, high level function. However, considerable advantage can be gained by distributing the decision mechanism.

The design of a multi-party decision mechanism for a decentralized system is not a trivial task because the various parties to the decision do not know the state of the other processes and are not necessarily synchronized. If two or more processes are involved in a transaction, or a set of concurrent transactions, then it may be necessary to use concurrency control techniques to ensure that each transaction is atomic and the results of a set of concurrent transactions are consistent (i.e. the same as if the transactions were executed sequentially) [8, 22].

For example, two phase locking protocols can be used to ensure the serialization of a transaction [23]. However, this protocol does not include a recovery mechanism and is vulnerable to communications or processor failure. Therefore two-phase commit protocols or non-blocking three-phase commit protocols are used to ensure that either all processes perform their part of a transaction or no process performs any part of the transaction [24, 25], this maintains the integrity of the decision process and the distributed database. A two phase commit protocol can be enhanced by including a timeout mechanism to prevent deadlock due to communications failures [21]. These protocols can be constructed using synchronous communication primitives and appropriate non-deadlocking transaction-abortion mechanisms and can be implemented easily on microprocessor based systems.

2.6 SOFTWARE FAULT TOLERANCE IN SEQUENTIAL SYSTEMS

The problem of designing fault tolerant software for sequential systems is much simpler than for concurrent systems, because error migration in the sequential system is bound by the lack of inter-process communication channels. This makes it easier to design appropriate error detection and recovery mechanisms such as recovery blocks.

The recovery block is a fundamental structure for fault tolerance in sequential systems. A design notation for the recovery block scheme is given below [4]:

```
ensure acceptance test AT
   by process P
   else by process Q
   else by default;
```

Before entry to the block, a copy of the system state is saved. The entry state is assumed to be error free. The process P within the recovery block is then executed and the results from process P are assessed using the acceptance-test AT. If the results are acceptable, the system is assumed to be error free and execution passes out of the recovery block. If the results from P are unacceptable, the system backtracks, restores the entry state, and recomputes the task using the diverse process Q. Normally the results from Q will be acceptable. However, it is necessary to include a default process which always generates acceptable results to ensure that the block terminates.

(a) Sequential real-time systems

The performance of a time-critical process can be monitored using a watchdog timer to ensure that the system complies with its timing requirements. The timer is preset to trip after a pre-determined time and is initiated to run concurrently with the time-critical process such that the first process to complete causes the other to abort. A watchdog mechanism can be incorporated naturally into the ensure notation described above. For example, the time-critical construct shown below [25] is designed to produce results which are both acceptable and timely:

```
ensure acceptance test AT
  within time t
    by primary process P
    else by default process;
```

Several strategies have been proposed for the sequential implementation of watchdog timers [26, 27]. Most require the use of a centralized scheduler which ensures that either process P or the default process completes before the deadline. In sequential systems it is usual for the process P to be executed first, although default-first algorithms have been implemented. The default-first strategy has the advantage of ensuring the availability of acceptable results, before attempting P. However, this strategy can also be computationally inefficient since the default results will be normally discarded as P completes successfully.

It is common practice to configure the watchdog timer to raise an external interrupt which interrupts the processor and causes the scheduler to abort the timed-out process P. However, the use of this technique results in a non-deterministic design. An alternative deterministic approach requires the process P to poll the timer to confirm that the timeout has not elapsed, this method assumes that P remains active. Care must be exercised in the design of the watchdog timer, particularly in applications which have implications for safety, where the form of watchdog may be prescribed by the relevant codes of practice.

Recovery blocks have been used in conjunction with watchdog timers to provide fault tolerance in real-time systems. This technique has been used also to guard against certain classes of transient hardware faults [28].

2.7 SOFTWARE FAULT TOLERANCE IN CONCURRENT SYSTEMS

Software fault tolerance for a concurrent system may be provided by suitably extending the recovery block technique. In particular, the error detection mechanism for a concurrent system must take into account the promulgation of errors through interprocessor communications in its assessment of error damage. Also, the error recovery scheme must involve all processes which may have been affected by the error. It follows that the structure of the inter-process actions in a concurrent system is central to the design of the fault tolerant mechanism.

Design methods for concurrent error detection and recovery schemes involve partitioning the set of concurrent processes into groups of processes, such that the processes within each group form a co-ordinated recovery block. The partition boundary must limit the propagation of errors which arise within the boundary, this will be achieved if the partition takes into account the structure of the inter-process communications. The backtracking recovery operation must be a properly co-ordinated procedure involving all the concurrent processes which are party to an error and must limit the extent of backtracking under fault conditions. Unless the recovery is carefully controlled, and its scope is bound, the action of recovery may extend beyond a proposed recovery block and may lead to progressive collapse of the whole system (the domino effect) [4].

The conversation mechanism proposed by Randell [4] uses a particular form of closed boundary to define a recovery block for a general set of distributed processes. Identifying the boundary of a conversation for a set of communicating concurrent processes is not a trivial task because the state of each asynchronous process is independent of the state of other processes, except when the processes are forced into synchronism by interprocess communications. In general, it is not possible to determine *a priori* the particular sequence of states which will be instantiated when the software is executed. Therefore, the boundary must be independent of the sequence of occurrence of the states or must be identified dynamically as the software is executed.

The boundary of the conversation consists of a recovery line, a test line and two side walls, which enclose the set of interacting processes which are party to the conversation. The recovery line defines the start of the conversation, it consists of a co-ordinated set of states (recovery points), one for each participating process. The participating processes enter the conversation asynchronously and each process saves its entry state for later use should recovery be required. The sidewalls isolate the body of the conversation from other processes and no communications are allowed through these walls; this prevents any errors generated within the conversation being propagated to other processes. The test line is a co-ordinated set of acceptance tests for the set of interacting processes. If the test is successful, the conversation is terminated and all processes exit synchronously. However, if the acceptance test is failed, recovery must take place. Recovery is achieved by rolling back the whole conversation to the recovery line, restoring the state of each process to the saved entry state, and executing a pre-defined alternative process or the default process. If the results from the alternative process are acceptable the conversation will terminate, otherwise the results from the default process will be used to exit the conversation. Thus all processes in the conversation co-operate in error detection and all participate in any subsequent recovery.

Conversations fault tolerant structures may be built into a system as a "static" entity during the design process, or may be created dynamically following an error. Static conversations may be generated by defining the boundary of the conversation early in the design process such that the conversation boundary will govern the design of the constituent processes and the inter-process actions (i.e. the dialogue between the processes). Alternatively, the boundary of a conversation may be placed in position late in the design process to protects a specific process or function.

A method has been developed [29] which allows the systematic placement of conversations to protect a specific process or to protect all the processes associated with a particular function.The technique is based on the analysis, using state-space methods, of the inter-process actions associated with a specified process or function. The aim is to reveal the underlying structure of the inter-process communications and identify a conversation boundary which bounds the extent of error migration. The method can be automated (using the underlying set theory) and provides a systematic method for the placement of conversations, or properly nested sets of conversations. The method can also be used in the verification of conversations which have been developed by direct synthesis or dialogues.

The problem of the dynamic identification of conversation boundaries has been the subject of much research [30, 31]. Considerable care is necessary in the design of dynamic recovery mechanisms because they may exhibit the domino effect. Therefore it is necessary to show that a specific dynamic recovery method is free from the domino effect before it can be used in a design.

(a) The implementation of conversations

Conversation mechanisms can be implemented by extending the ensure notation described in Section 2.5 to provide a structure for concurrent systems. The general structure of the conversation will be similar to that of the sequential recovery block, however the structure can be simplified by initiating and computing the primary and alternative processes in parallel using identical copies of the initial high integrity system states. Thus, there will be no need to establish a recovery point or a process roll-back mechanism. The structure of the acceptance test requires careful design and the acceptance test must be executed in the precise sequence defined in the ensure notation [32]. In general, the acceptance test can be performed in two phases comprising local tests, which are are carried out on each constituent process in the conversation, and global acceptance test which assesses the complete set of results from the conversation. If the results pass both the local and global acceptance tests then the conversation is successful and the constituent processes should exit synchronously from the conversation. However, if the results of any process fail the acceptance tests, at either local or global levels, the results are discarded in their entirety and the results of the alternative processes are used. In both cases, the conversation must signal unsuccessful processes to abort in a deadlock free manner [33]. The test-line process and recovery mechanism can be implemented as a centralized process. Alternatively, the decision process may be distributed among the constituent processes and implemented using multi-phase commit protocols (or structured dialogues) [34, 35].

(b) Concurrent real-time systems

A concurrent timeout mechanism can be designed in which the watchdog timer is initiated and executed in parallel with both the primary and default processes [21], such that process P and the timer race asynchronously against each other as shown below:

```
PAR
  P
  timer
  default
```

Considerable care is required in the detailed design of a concurrent timeout mechanism if the design is to be deterministic and free from deadlock. In particular, the following cases must be accommodated without fault or deadlock: P completes first; timer completes first; P and timer complete simultaneously; process P is "late" but still active; P hangs and is unresponsive to communications commanding it to abort. A solution is to use a co-ordinating process which incorporates two parallel subprocesses to handle process termination and abortion. One subprocess is activated primarily by the completion of P and will abort the timer; the other is activated primarily by a timeout trip and aborts P. Both subprocesses are guaranteed to run to completion because the inactive subprocess is activated by the process which aborts. Petri net analysis shows that this design is free from deadlock and provides timeout protection, except in the case in which P hangs [21]. In the latter case, the design should ensure that the system can operate satisfactorily using the default process and that if (or when) deadlock occurs, its external effects are minimized.

(c) The protection of communications

Software fault tolerance techniques for concurrent real-time systems can also be used to protect the inter-process communciations which ensure the proper synchronization and operation of a distributed system [19, 20]. This is particularly important because these systems depend on the integrity of the inter-process communications.

Communication transactions can be protected using watchdog timers only if there is a logical pairing of the protective mechanisms on each side of the transaction; this often requires an extension to conventional concurrent programming languages. Alternatively, a solution can be obtained by decreasing the level of granularity at which protection is applied, by enclosing a communication which must be protected within a conversation. One property of the conversation is the absence of communications through the sidewalls. Therefore, it is possible to protect the conversation using a watchdog timer which monitors the performance of the complete set of processes and communication transactions within the conversation. In the event of a participating process becoming unresponsive, or a failure in the communications between processes, the conversation will not terminate and the default process or conversation will then be invoked by the timer. This type of design can be implemented using conventional concurrent programming languages.

2.8 TRANSACTION PROCESSING AND DISTRIBUTED DECISION MECHANISMS IN DISTRIBUTED SYSTEMS

The problem of transaction synchronization can be solved using concurrency control techniques [8, 22] which maintain the integrity of a decision process by ensuring that the result of a decision process does not depend on whether the component transactions are executed concurrently or in some serial order. Locking is the the most popular and widely used method of ensuring serialization. It involves a transaction locking a resource so that any other transaction requiring access to the resource must wait until the original lock is released. To increase the availability of a resource, two modes of locking are used; "shared" locking is used in transactions that only want read-only access to a resource and "exclusive" locking, which guarantees mutual exclusion, is used in any transaction that wants to alter the state of the resource. Concurrency control techniques such as two phase locking inevitably reduce concurrency as processes wait for resources. The reduction in concurrency depends on the granularity of the resources being locked. The two phase locking technique, which guarantees serializability, has been proved to be a correct locking algorithm for centralized databases [24] and distributed databases [8]. Two phase locking protocols do not include a mechanism for recovery after a failure. For example, deadlocks may occur if a transaction holding a lock on an item fails, thus forcing other transactions which require the resource to wait indefinitely. This problem is usually solved by using a timeout mechanism.

In a distributed system care is required to ensure that the database is consistent. Distributed databases are more prone to failure than centralized databases due to individual site failure or communication failure. For example, problems may arise when more than one copy of a data item is stored at independent sites because one site may be updated while another may fail to achieve the update and the database will become inconsistent. This inconsistency can be removed by using protocols that ensure the all or nothing property (i.e. atomicity) of transactions.

Two-phase commit protocols, or non-blocking three-phase commit protocols can be used to ensure that either all processes perform their part of a transaction or no process performs any part of the transaction [8,24]. The two phase commit protocol defines a co-ordination process which controls the participant transactions, acts as a judicator and gives a unique decision. The decision process comprises a two phase protocol. In the first phase, the co-ordinator asks the participants if they are ready to commit. If a participant is ready to commit then a ready message is returned otherwise an abort message is sent back. In the second phase, the co-ordinator receives messages from all participants and makes the decision to commit or abort. If all participants reply ready to commit, then they are all commited and the transaction is completed. However, if any participant returns an abort message then all participants not already aborted are aborted. This ensures that all transactions reach the same decision and are informed of the decision explicitly.

These protocols can be constructed using synchronous communication primitives and appropriate non-deadlocking transaction-abortion mechanisms [36]. Also, a two phase commit protocol can recover safely, even after a processor failure, using a log record kept in "safe storage" for the co-ordinator. Typically, a two phase commit protocol is used to ensure the integrity and recoverability of the decision process [24, 25]; this type of protocol can be implemented easily on microprocessor based systems.

In a distributed system, a failure in the communications network may result in the loss of messages, such as those passed between the co-ordinator and the participants in a two phase commit protocol. To prevent deadlock the two phase commit protocol can be enhanced by including a timeout mechanism which is triggered if answer messages from the participants (either commit or abort from each participant) are not received within the timeout period. In the event of a timeout, an abort command message would then be sent to all participants. Since the first phase of the commit process and the timer process are made to race asynchronously, provision has to be made to absorb outstanding communications from the process which loses the race by arranging for the co-ordinating process to abort the losing process [21].

2.9 CONCLUSIONS

This chapter has discussed the problem of providing software fault tolerance in real-time systems and has discussed the principal techniques involved. It has shown that there is a role for software fault tolerance techniques in the design of reliable systems.

In particular, software fault tolerance techniques for sequential and concurrent real time systems, such as recovery blocks, conversation schemes and concurrent watchdog mechanisms, can be used to protect specific processes and functions. They can also be used to protect the interprocess communications which ensure the proper synchronization and operation of distributed systems. Many of these techniques have been proved in practice.

The newer concurrent techniques, such as distributed decision mechanisms, will undoubtably prove useful in the design of robust concurrent systems, such as real-time control systems. The need for these techniques will grow due to the sophistication of the logic structures, and the subtlety of the bugs, in advanced concurrent real-time systems.

REFERENCES

1 Musa, J., Iannino, A. and Okumoto, K., "Software reliability", McGraw Hill, 1987.

2 Anderson, T., and Lee, P.A., "Fault Tolerance, Principles and Practice", Prentice Hall, 1981.

3 Avienzis, A., "The N version approach to fault tolerant software", IEEE SE, Vol. SE-11, No. 12, pp. 1491 - 1501, 1985.

4 Randell, B., "System structure for software fault tolerance", IEEE Trans. SE,Vol. SE-1, pp. 220-232, 1975.

5 Hecht, H.,"Fault tolerant software for real-time applications", ACM Computing Surveys, Vol. 8, No. 4, pp. 391-407, 1976.

6 Young, S.J., "Real-time languages: design and development", Ellis Horwood, 1982.

7 Inmos., "Occam programming manual", Prentice Hall, 1984.

8 Ceri, S. and Pelagatti, G., "Distributed database principles and systems", McGraw-Hill, 1984.

9 Korth, H.F., and Siberschatz, A., "Database system concepts", McGraw Hill, 1986.

10 Hoare, C. A. R., "Communicating sequential processes" , Prentice Hall, 1986.

11 Peterson, J.L., "Petri net theory and the modeling of systems", Prentice Hall, 1981.

12 Mekly, L.J., and Yau, S.S., "Software design representation using abstract process networks", IEEE Trans. SE, SE-6, pp. 420-434, 1980.

13 Hecht, H., "Fault tolerant software", IEEE Trans. on Reliability, R-28, pp. 227-232, 1979.

14 Anderson, T. (Ed), "Resilient computing systems", Collins Professional and Technical Books, 1987.

15 Holding, D.J. and Carpenter, G.F., "Software fault tolerance in real-time systems", Ch 8. in "Parallel processing in control – the transputer and other architectures", P.J. Fleming (Ed), Peter Peregrinus Ltd., London, 1988.

16 Knight, J.C., and Levenson, N.G., "An empirical study of failure probability in multiversion software", Proc. 16th Int. Symposium on Fault Tolerant Computing Systems, pp. 165-170, 1986.

17 Levenson, N.G., "Software Fault Tolerance; the case for forward error recovery", Proc. AIAA Conf. on Computers in Aerospace, pp. 50-54, 1983.

18 Campbell, R.H., and Randell, B., "Error recovery in asynchronous systems", IEEE Trans. Software Engineering, SE-12, pp. 811-826, 1986.

19 Kramer, J., Magee, J. and Sloman, M., "Intertask communication primitive for distributed computer control systems", Proc. 2nd Int. Conf. on Distributed Computer Systems, Paris, pp. 404-411, 1981.

20 Holding, D.J., Carpenter, G.F., and Tyrrell, A.M., "Aspects of software engineering for systems with safety implications", Proc. 6th IEEE/Eurel Conf. on Computers in communications and control (Eurocon 84), Brighton, England, pp. 235-239, 1984.

21 Carpenter, G.F., Holding, D.J., and Tyrrell, A.M., "Analysis and protection of interprocess communications in real-time systems", Journal IERE, Vol. 58, No. 4, June 1988.

22 Bernstein, P.A. and Goodman, N.,"Concurrency control in distributed database systems", ACM Computing Surveys, Vol. 13, No. 2, pp. 185-221, June 1981.

23 Erswaren, P.K., Gray, J.N., Lorie, R.A. and Traiger, I.L., "The notions of consistency and predicate locks in a database system", Comm. ACM, Vol. 19, No. 11, pp. 624-633, Nov. 1976.

24 Gray, J.N., "Notes on database operating systems", in Lecture notes in Computer Science, Vol. 60, pp. 393-481, Springer-Verlag, 1978.

25 Balter, R., "Selection of a commitment and recovery mechanism for a distributed transactional system", IEEE Proc. 1st Symp. on Reliability in distributed software database systems, pp. 21-26, 1981.

26 Campbell, R.H., Horton, K. and Belford, G.G., "Simulations of a fault tolerant deadline mechanism" in Digest of papers, Fault Tolerant Computing Systems, Madison, pp. 95-101, 1979.

27 Upadhyaya, J.S. and Saluja, K.K., "A watchdog processor based general roll back technique with multiple retries", IEEE Trans. Software Engineering, SE-12, pp. 87-95, 1986.

28 Jackson, P.R., and White, B.A., "The application of fault tolerant techniques to a real-time system", Proc. Int. Conf. on Safety of Computer Control Systems (Safecomp '83), pp. 75-82, 1983.

29 Tyrrell, A.M. and Holding, D.J.,"Design of reliable software in distributed systems using the conversation scheme", IEEE Trans. on Sofware Engineering, SE-12, pp. 921-928, 1986.

30 Merlin, P.M., and Randell, B., "State restoration in distributed systems", Proc. 8th Int. Symp. on Fault Tolerant Computers, pp. 129-134, 1978.

31 Russell, D.L., "State restoration in systems of communicating processes" IEEE Trans. Software Engineering, Vol. SE-6, pp. 183-194, 1980.

32 Carpenter, G.F., Holding, D.J., and Tyrrell, A.M., "The design and simulation of software fault tolerant mechanisms for application in distributed processing systems", Microprocessing and Microprogramming, No. 22, pp. 175-185, May 1988.

33 Carpenter, G.F., "The use of Occam and Petri nets in the simulation of logic structures for the control of loosely coupled distributed systems", Proc. UKSC Conf. on Computer Simulation, Pub. SCS, pp. 30-35, Sept. 1987.

34 Campbell, R.H., Anderson,T., and Randell, B., "Practical fault tolerant software for asynchronous systems", Proc. Int. Conf. on Safety of Computer Control Systems (Safecomp '83), pp. 59-65, 1983.

35 Gregory, S.T. and Knight, J.C., "A new linguistic approach to background error recovery", Proc. 15th Int. Symp. on Fault Tolerant Computing, pp. 404-409, 1985.

36 Holding, D.J., Hill, M.R. and Carpenter, G.F., "The design of distributed software fault tolerant, real-time systems incorporating decision mechanisms", Proc. 14th Symp. on Microprocessors and Microprogramming – Euromicro 88, Zurich, Sept. 1988.

Chapter 3

SOFTWARE RELIABILITY: THE WAY FORWARD

D.J. Smith
Technis

3.1 ADVANTAGES AND DISADVANTAGES OF PROGRAMMABLE SYSTEMS

A programmable system is any equipment or device which, having a computer architecture relies on a set of sequential programmed instructions in order to function. This set of logical commands is referred to as Software and the term also embraces the design documents as well as the code.

For both reliability and operating reasons there are advantages and disadvantages arising from programmable devices.

(a) Advantages

- Less hardware (fewer devices) per circuit due to high levels of integration.
- Fewer device types.
- Consistent architecture (configuration) leading to a common approach to hardware design.
- Easier to support several models in the field with spares.
- Simpler to modify or reconfigure.
- Provides a running log for investigation.
- Allows for self test and early warning diagnostics.

- Removes human operators from hazardous areas and provides sophisticated process interlocks.
- Interprets rates of change of process parameters and gives timely warning of potentially hazardous conditions.
- Provides centralized displays and graphics on VDUs.

(b) Disadvantages

- Difficult to "inspect" software for inherent faults.
- Difficult to impose standard approaches to software design.
- Difficult to control Software changes.
- Testing and validation of high scale integration devices and their associated software is difficult owing to the high package density and consequent lack of visibility and interface to the functions.
- Impossible to predict software failure modes and rates.
- Exhaustive testing is impossible because of time constraints since the number of permutations in software is extremely high.
- More susceptible to Common Cause Failures whereby a single failure causes loss of redundant protection.
- Easier to corrupt data and programs.

3.2 FEATURES OF SOFTWARE RELATED FAILURES

(a) Terms

Unlike hardware failures there is no physical change which causes a unit to cease functioning. Software failures are errors which, owing to the complexity of code, seldom become evident immediately. Unlike the hardware Bathtub Curve, there is no Wearout feature since the population of "bugs" can only (save for modifications) decrease. Figure 3.1 illustrates the concept of FAULT/ERROR/FAILURE.

FAULTS may occur in both Hardware and Software. Software faults – often known as BUGS – will arise as a result of particular parts of the code being used for the first time or because of corruption due to some outside influence.

The presence of a fault in a program does not necessarily result in either error or failure. A long time may elapse before that portion of the code is used under circumstances which lead to failure.

A FAULT (bug) may lead to an ERROR. This is the condition whereby the system is in an incorrect state. A data value or an instruction thus incorrect remain so until the particular part of the code is executed which reveals the ERROR.

An ERROR may propagate to become a FAILURE if the system does not contain some error recovery logic capable of dealing with a minimizing the effect of the error.

A FAILURE, be it hardware or software related, is the termination of the ability of an item to perform its specified function.

(b) Causes of faults

Faults arise from all stages in the design process. There is evidence that the majority of errors (over 60%) are committed during the requirements and design phases. The remaining 40% occur during coding. That is not to say that coding is not a part of the design but it is only the final activity in a much larger process. The more complex the system the more faults will be likely to stem from ambiguities and omissions in the specification stages. The major sources of fault are:

(a) The Requirement Specification

- Incorrect requirement due to:

    Model not a good fit to the physical situation.

    - Incorrect document cross references.

- Inconsistent or incompatible requirements:

- Two references give conflicting information.

- Requirement unclear or illogical.
- Requirement omitted:
  - e.g. handling of invalid inputs.

(b) The Design

- Unstructured approach to the design breakdown (i.e. detail is considered first).
- Use of non-standard language.
- Lack of change control.
- Specification was misunderstood.

(c) Coding

- Semantic errors involving incorrect use of statements.
- Logical errors in translating the design into code.
- Detailed syntax errors which may have escaped detection by the compiler.
- Use of an incorrect condition at a branch.
- Poor data validation (e.g. no default condition after a data input)
- Variables not initialized, or used incorrectly.
- Insufficient arithmetical accuracy.
- Insufficient range checks (e.g. divide by zero).
- Type mismatch (e.g. string used as a variable).
- Residual errors in compilers.

(c) Modelling

It must be emphasized that software failures are not time related since they are revealed as a result of exercising new paths in the software in conjunction with the combination of real time inputs to the system. At present, attempts to model the distribution of failures are hampered by lack of adequate field data to validate them.

(d) Some typical safety system faults

In the course of carrying out assessments some common faults have been noted and have led to timely improvements in both the operating procedures and the design.

Design

- Lack of functional diversity whereby safety functions can be lost due to a single software failure.
- Programmable systems in series whereby a software derived safety signal becomes the input to a second programmable safety system.
- Inadequate document structure (Figure 3.1)
- Scope for additional watchdog or error detection facilities.
- Inadequate provision for extension which ultimately leads to degraded functions when the system is extended.

Operation

- Inadequate Control of Changes.
- Operating documentation fails to address procedures under default conditions.
- Maintenance routines fail to recognize the importance of testing for common cause failures in redundant system.
- Inadequate control of media.

3.3 RELIABILITY/INTEGRITY IN DESIGN

(a) The meaning of "Design Cycle"

The idea of a software Design-Cycle (Figure 3.2) is a convenient model which serves two purposes. Firstly it allows one to represent the process of conception and production in a graphical and logical form and secondly, it provides a framework around which the assurance activities can be built in a disciplined manner.

The important feature of Figure 3.2 is the existence of the loops which represent a review or test at each stage of the design. Faults discovered earlier in the cycle will cost less to remedy and are less likely to propagate into field use.

(b) Configuration and diversity

Redundancy is often used as a means of improving system reliability. Peripheral signals are fed to two (or more) channels or processors.

If the channels do not agree then processing is not allowed to proceed since there is no determination of which channel is healthy. An alarm or shutdown condition could then be initiated. In this case the overall reliability may be less than for a single channel since twice the number of failures will be likely. On the other hand safety is enhanced since, for a hazardous failure mode, both channels need to fail. Variations of this principle involve self test and two out of three voting.

However, any software fault, with simple hardware redundancy, will be a Common Cause Failure since the identical code will exist in each equipment. Recent studies have shown that the number of Common Cause Failures due to software in real time redundant systems is greater than for other equipment.

One approach to this problem involves **SOFTWARE DIVERSITY** which is a particular form of redundancy involving the design and coding of separate Software for each of the replicated channels. Sometimes called **N Version Programming,** this is not only expensive but is not a total defence since specification related faults will propagate through each of the N designs.

(c) Fault tolerant designs

A number of Software features can be included which improve the integrity of systems under fault conditions. These include:

Fault identification by use of :

- Watchdog timers
- Cyclic memory checks
- Relay runner techniques
- Built in tests
- Range and variable checks

Error correction by means of :

- Parity
- Checksums
- Reinitialization at known state
- Recovery blocks
- Exception handling

(d) Graceful degradation and recovery

The overall design philosophy should take account of the need to operate in degraded modes. Functions should be partitioned so that single failures cause only degraded performance rather than loss of the total safety functions. This can only be achieved if taken into account by the requirements where functional diversity can be specified, levels of function can be defined and operating requirements grouped into categories.

The Software design should attempt to minimize the routes of communication between groups of modules so that errors are discouraged from propagating. Thus errors are more likely to be confined to single functions. The system may then be able to provide service, albeit at a degraded level, by means of other functions. This is of particular importance in controlling hazardous processes.

3.4 THE APPROACH TO SAFETY/RELIABILITY ASSESSMENT

(a) Defining the safety system

The first step is to define the failure modes/hazards to be considered. The results of an assessment may well be different for each failure.

(b) Assess the hardware and the overall configuration

Initially the hardware reliability should be assessed using one of the current techniques (i.e. Fault Free Analysis, FMEA). Particular attention should be given to the possibility of Common Cause Failures.

(c) Software quality review

This involves a detailed review of:

- Document Structure
- Document and Media Control
- Design Standards
- Coding Methods
- Languages and Compilers
- Design Review Records and Methods
- Validation Methods
- Test Strategy
- Test Records
- Records of Remedial Action

This is described in detail in reference 1 (HSE Document "Programmable Electronic Systems in Safety Related Applications", 1987) which includes extensive checklists to assist in reviewing the above areas.

Software quality assessment is also described in reference 2 ("Engineering Quality Software", Elsevier Applied Science, 1987) together with a review of the new techniques being developed.

3.5 LIMITATIONS AND DRAWBACKS OF SOFTWARE ASSESSMENTS

(a) Addressing relevant parameters

In most cases, assessments address those aspects of Software design which are covered by traditional QA, as listed in 3.4(c).

Inevitably the questions which are addressed in each of these areas tend to be of a generic nature and, as a consequence, do not permit conclusive answers. Indeed, both positive and negative responses can be justified from a single question as, for example:

"Have documentation and coding standards been adequately addressed ?"

From one point of view the answer can be "Yes" in that some reasonable level of guidelines has been provided. An equally justifiable "No" can be suggested in that the coding and design standards cannot guarantee the absence of faults. This open ended qualitative checklist approach implies a lack of precision making software quality impossible to assure.

Whilst such controls and reviews eliminate many Software related faults it must be realized that they operate at a relatively superficial level with respect to those factors, in the design and code, which ultimately create errors.

(b) Error rate versus fault tolerance

A frequent misconception is that the elimination of faults is the sole factor in achieving quality Software. This is too simplistic an assumption. In practice all software is likely to contain residual faults albeit at very low levels even after extensive QA and debug activities. Compare, therefore, two safety systems one of which has a few unknown residual faults and the other which has twice as many.

Assume that the code having the greater number of faults has been carefully structured and limited in its ability to access variables and codes in such a way as to restrict the propagation of errors. Assume, also, that there are a number of reinitializations at known acceptable values when an error is detected.

If, in addition to these features, the safety system and its software is designed in such a way that individual failures do not cause total loss of function then it will at least continue to offer a degraded level of protection. An example of the latter would be a fire protection system which measures more than one parameter (i.e. U.V. light, smoke, rate of temperature rise). The interpretation of each type of input, and the generation of executive output action, could be dealt with by separate parts of hardware and software.

This fault tolerant type of design offer a far higher level of integrity then the system with less faults and far worse consequences in the event of failure.

(c) False confidence

Despite the above shortcomings, there is a tendency to assume that because a safety system has been assessed then the mere fact of assessment has somehow "qualified" the software. It does not recognize that inherent faults will almost certainly exist and that the assessment process did not admit the precision necessary to reveal them.

3.6 THE WAY FORWARD

(a) Formal expression of requirements

The drawbacks outlined in Section 3.5 mostly relate to the problems of ambiguity inherent in free language. Nevertheless, despite considerable advances in the formality and checking capabilities of coding languages, requirements are still written in free language text. It has long been recognized that this is a major limitation when pursuing the objective of fault free software. Many attempts have been made to develop formal requirements languages. They are, however, highly structured and supported by graphical methods. IORL, CORE, VDM, Z and OBJ are the names of such methods.

MASCOT, SSADM, JSD are examples of similar methods applied to the design specifications.

(b) Static analysis

This involves the use of automated software tools which consist of a number of analysis routines. Typically these include:

- **Control Flow Analysers**

  These identify all possible starts and ends, unreachable code and "black holes". It gives an initial "feel" for the quality of the program. If this not good then there is every likelihood that subsequent analysis will not result in a good program.

- **Data Use Analysers**

  These identify all the inputs and outputs and checks that data is not being incorrectly handled (e.g. read before it has been written).

- **Information Flow Analysers**

  Outputs are analysed to describe which inputs they depend on (e.g. output Z depends on inputs A, B, C).

- **Partial Program Generators**

  These extract subprograms which cater for particular variables of interest. This helps to reduce complexity. These subprograms can then be submitted to the semantic analyser.

- **Semantic Analysers**

  These provide the functional relationships between variables (e.g. P depends on A*Bsq + C).

  They identify what the program is doing for each path and thus provides a means of assessing whether the program meets the specification.

- **Compliance Analysers**

  These take the output from the semantic analyser and compares it with an embedded specification. For example, if X must be in the range -3 to $+90$, the analyser tests to see if the condition is met, in the program.

Available analysers at present are MALPAS (Rex Thompson and Partners, Farnham, Surrey), SPADE (Southampton University) and LDRA (Liverpool University).

(c) Test beds

The test process lends itself to automation and much effort is being invested in the development of automated test beds and test animators. These involve automating the application of test stimuli as well as displaying a variety of data for interpretation during the tests.

Examples include:

- **Drivers**

  These provide a means of simulating inputs.

- **Test beds**

  Provide simulations and also a means of displaying data, variables and executed code as the test progresses.

- **Emulators**

  Provide an "environment" which simulates responses as well as inputs.

- **Analysers**

  Test alternate paths, executive software at extreme values and seed incorrect values.

(d) Conclusion

The increased use of programmable devices as safety systems, together with an increased requirement for integrity assessment, emphasizes the need for these Software quality techniques.

Provided that the limitations and pitfalls, described in this chapter, are borne in mind then the Software assessment provides a valuable additional review within the design cycle.

The formal tools, described briefly in Section 3.6, will considerably enhance the reliability of Software products as well as providing a further step towards being able to certify software.

REFERENCES

1 "Guidance on the use of programmable electronic systems in safety related applications", Health and Safety Executive, 1987.

2 Smith, D.J., and Wood, K.B., "Engineering Quality Software", Elsevier Applied Science, 1987.

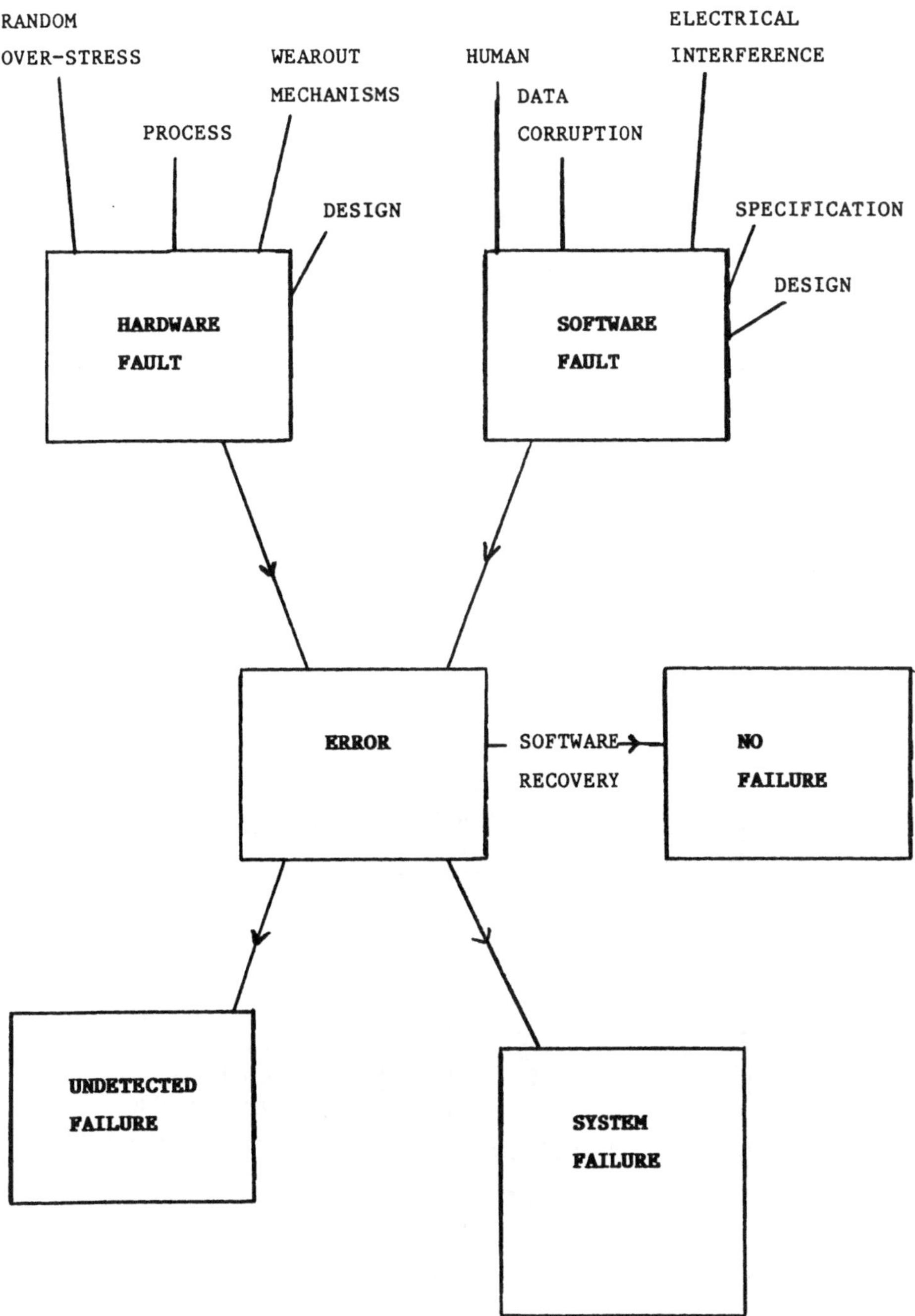

Figure 3.1. Fault/Error/Failure.

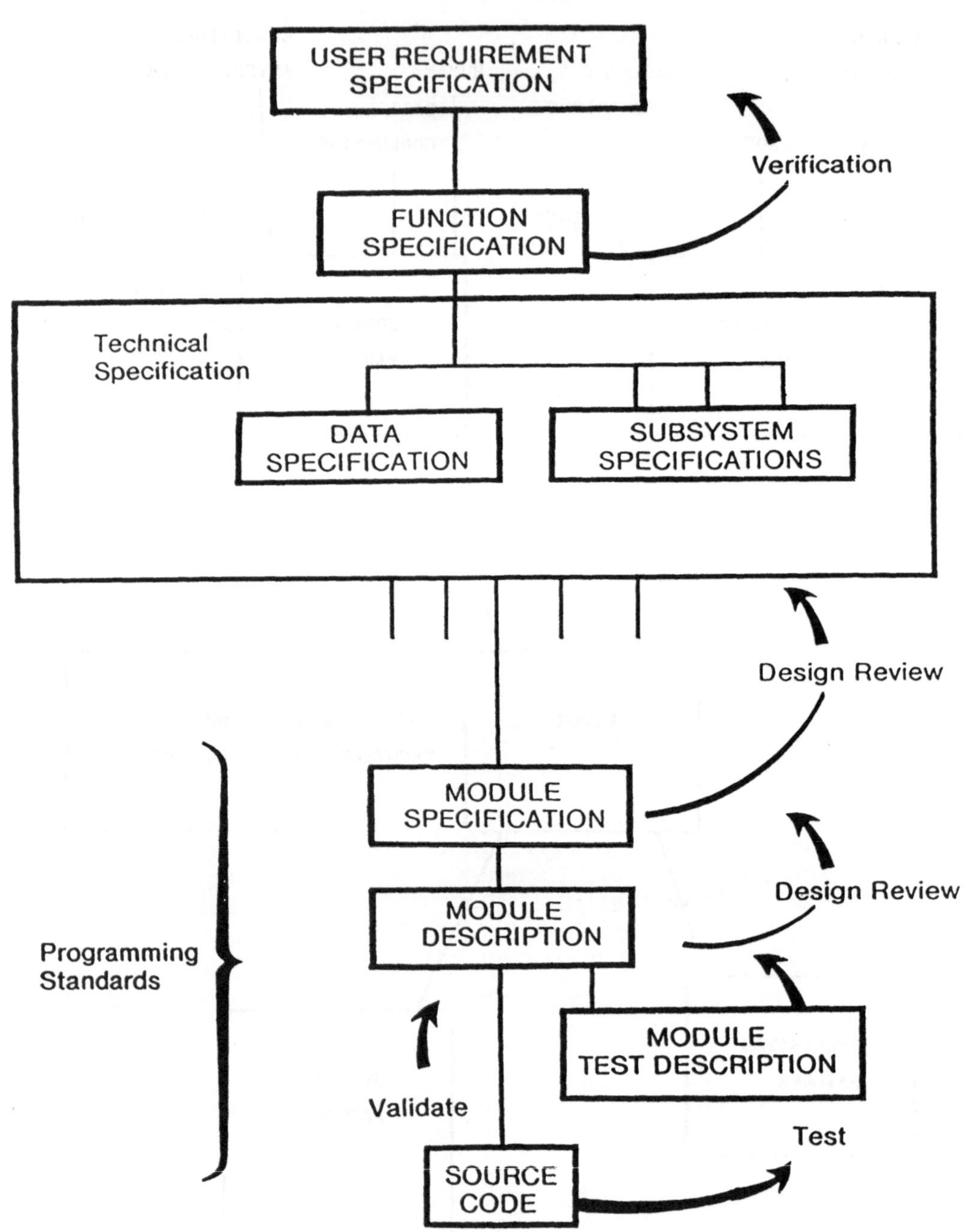

Figure 3.2. The Design Cycle.

Chapter 4

DESIGN PRINCIPLES FOR DIAGNOSTICS DECISION SUPPORT SYSTEMS

D.E. Embrey and T.G. Wolfmaier
Human Reliability Associates Ltd.

4.1 INTRODUCTION

This chapter is intended to provide an overview of some of the fundamental design principles that need to be taken into account in the development of a diagnostic support system. Important concepts such as the mental model of the operator are described, and the role of information acquisition and training in the development of a diagnostic DSS are discussed.

4.2 THE OPERATORS' MODEL

Despite developments in computer based control, the most important component in the operation of systems is still the operator. In order to control a process or troubleshoot a failure the operator requires an internal mental model of the system. This mental model enables him:

- To sample and interpret information displays.
- To plan interventions.
- To compare the actual state of the system as indicated by information displays with an expected state derived from the mental model. This provides a basis for fault detection when the actual state does not correspond with the expected state,

as well as fault diagnosis to the extent that the quality of the discrepancy indicates the cause of the failure.

The internal model is specific to individual systems. It contains information about:

- Form and location of components of the system.
- Functions of these components of the system.
- Causal relations between components and system parameters, such as the relation between valve position and pressure.

4.3 DIAGNOSTIC SUPPORT

The system is equipped with sensors which monitor certain critical parameters, such as temperatures, pressures, valve positions, etc. in parts of the system. If one or more of these parameters reach a critical state, i.e. increase over or decrease under a critical value, then an annunciator is activated. The operator has to identify the cause of the failure and restore normal functioning as quickly as possible.

Three main methods of improving the diagnostic performance of operators exist:

- Providing operators with **information displays** to monitor relevant system parameters.
- Providing a **diagnostic support system** to assist them in their diagnostic activity.
- Providing operators with **training** to support diagnosis.

(a) Information displays

The failure of a system is rarely a sudden event. It is preceded by changes in system parameters until they reach the critical point that finally triggers the alarm. The characteristics of these changes in system parameters prior to an alarm, the system parameters involved and the dynamic of the changes, provide substantial diagnostic information about the event and the cause of a failure even before it actually occurs.

Therefore, providing operators with an information display to monitor relevant system parameters has two advantages:

- It facilitates early fault detection before a critical state has been reached and an alarm has been activated. This enables the operator to intervene at an early point in order to prevent system shutdown and maintain optimal functioning.
- It improves fault diagnosis by providing the operator with diagnostic information about the cause of the failure based on the characteristics of the initial changes in system parameters. These improvements in diagnosis will reduce shutdown time.

In order to benefit from the information display it is important that the information is presented in a way that matches the mental model of the operators. The greater the correspondence between the form of the information display and the operators' model, the easier, quicker, and more reliable the processing of the information and the decisions based on this information.

(b) Diagnostic support system

When a failure has been detected either by the operator or by the annunciator system, the operator has to identify the fault and restore normal functioning.

In this context the operator could encounter two problems:

- The operator has an inappropriate or incomplete internal model of the system and is therefore either unable to identify the cause of the failure or arrives at wrong conclusions.
- The operator has an appropriate and complete internal model but due to stress arising from the diagnostic situation he focuses his attention on the first obvious explanation and neglects other alternatives. This so-called "tunnelling effect" provides a severe limitation to the operator's diagnostic performance.

As a consequence of this, the operator needs a diagnostic support system which provides him with the information to identify and repair the fault. He does not require information about the whole system. In most cases the failure is limited to a certain aspect of the system, and therefore only information related to the current failure is needed.

For this purpose rule-based diagnostic aids in the form of fault-symptom matrices or symptom-based procedures have been used. But they have several problems:

- They provide information in a textual form which makes it difficult for the operator to relate this information to his spatial mental model and the actual system.
- They do not enable the operator to cope with unpredicted events, i.e. failures that are not covered by the diagnostic guide.

More recently developed diagnostic support systems have the potential to overcome these problems. Basically, they provide the operator with three types of information:

- A spatial-physical graphical representation containing information on how the components look and where they are located. This is particularly important for the trouble shooting of mechanical faults. Ideally, this would include some kind of three-dimensional view of the system.
- A schematic representation of the components, containing information about their functions and their causal relation to the affected system parameter. This schematic display supports the operator in deducing possible causes.
- A list of common faults. This list does not have to be complete but if possible it should cover about 90% of all faults. This list accomplishes two functions:
  (a) If the operator is inexperienced this list of common faults, possibly ordered according to the frequency of their occurrence, could guide the operator's diagnostic activity in a manner similar to proceduralized diagnostic support systems. Additionally, the operator should have the option to retrieve more information about faults in the list, such as related symptoms, verification, and repair and adjustment procedures. Again this information has to be presented in correspondence to the operator's mental model of the system.

(b) If the operator is experienced the list could serve as a quick reminder of frequently occurring faults and thereby reduce the limiting effects of stress arising in the diagnostic situation.

It is important that this information is related only to the existing failure so that the operator can concentrate on essential information. Otherwise he will be confused and disturbed by the additional unnecessary information. This requires a knowledge-based diagnostic support system that adapts the information provided to the information required by the operator in the particular situation. For this purpose, the system has to be interactive, i.e. it has adapted the system's output to the operator's input. Such an interactive, knowledge-based system could be designed by using computers for the presentation of diagnostic information.

(c) Training

The main objective of training is to guide the operator in constructing an appropriate mental model of the system which enables him to interpret the information displayed, to plan interventions and to detect and diagnose faults. This construction of an internal model could be supported by an introduction which includes information on the physical aspects of components, such as where they are located, together with functional information on how they work. Operators also need causal information about how they causally determine system parameters, such as temperatures, pressures etc.

After the operator has acquired a initial mental model of the system, he has to learn how to apply this knowledge to controlling the process and troubleshooting failures. Thus, he can successively confirm, complete and refine his model. This could be achieved by simulation techniques. It has been shown that even the low-fidelity simulations can have substantial effects on improving operators' performance. But some kind of hands-on experience on the real system will also be required.

In order to benefit from information displays the operator requires training that enables him:

- To sample and interpret the information displayed and translate it into a model of the current state of the system.
- To derive discrepancies between the actual and the expected state of the system.
- To detect patterns in the changes of system parameters that precede certain failures or are related to certain faults.

This training is very important not least because if the operator lacks of confidence in his ability to predict a failure then he will be afraid of raising a false alarm and will wait until the fault has influenced the system to that extent that an annunciator is activated.

An important aspect of every training scheme is to ensure that the operators receive regular re-training. The extent to which re-training is required is a subject for investigation.

In the context of a diagnostic DSS, training has a critical role to play. The DSS must take into account the expected level of training of the operator, and provide the degree of support that is appropriate. Thus, the design of the DSS and the training system are closely interlinked. For this reason the DSS itself has a central role to play in training in that it embodies a considerable proportion of the knowledge necessary for successful diagnosis. It is also the "partner" with which the operator will interact whilst carrying out a diagnosis, and hence it is essential that a symbiotic relationship is developed between operator and DSS during training.

4.4 CONCLUSIONS

There is no single approach to developing operator support for diagnosis. Information displays, a diagnosis support system, and training are related and a balanced combination will be required to produce substantial improvements in diagnostic performance.

Chapter 5

A FAULT TOLERANT CONTROL SCHEME

K. Warwick
University of Reading

5.1 SUMMARY

A large number of computer based feedback controllers make use of the computer element simply as a numerical manipulation tool. At best a rudimentary jacketing arrangement is employed, on the lines of fail-safe switching when a measured level extends beyond a specified range. Hardware design has, of late, been directed largely towards a parallel implementation base [1] and it seems only sensible that controllers are redesigned to make use of this. A fault tolerant control scheme is described here, which is explicitly parallel in nature and therefore immediately suitable for parallel implementation, the overall procedure being directly dependent on the use of a hierarchically structured intelligent overseer.

5.2 INTRODUCTION

Before attempting to make a control system tolerant towards faults, it is important to answer the following questions :

- What is the nature of the fault?
- Is it critical?
- How is the performance affected?

Taking these questions in order, the first is highly dependent on where the fault occurs, that is there could be (i) a fault on the plant itself. The nature of the fault is then very much dependent on the nature of the plant itself, its complexity and its importance within an overall arrangement. In many cases plant faults can be dealt with by means of a knowledge base for comparison purposes, or more appropriately, via an expert system, thus intermediate plant signals can be monitored and compared with those expected. In the simplest case, if signals are lower (or higher) than expected then a sequence of pre-programmed events are brought into effect, on the lines of :

IF A THEN DO B

or **IF A THEN CHECK C TO SEE IF B IS NECESSARY**

The procedure for a plant fault is often to witness the fault effect and from this to try and establish the exact cause of the fault, or the most likely cause [2]. Subsequently a remedy or fault removal is carried out, as deemed appropriate.

All of these actions with regard to a plant fault are dependent on how well the plant has been modelled, i.e. how large the knowledge base pertaining to the plant is.

Rather than the plant being faulty, it could be that there is (ii) a sensor failure or measurement system error. The action to be taken in such cases is dependent on how much information is collected in order to pinpoint the error. A simple example is the creation of a multi-sensor environment, i.e. multiple path, in which all sensors read the same value; if one is in error this is easy to detect. Such a procedure though can prove extremely costly, and hence Intelligent Sensors are now becoming much more popular.

Although with this method each individual sensor is more expensive, the need for a multi-sensor bank is lessened. Self-checking and assessment techniques, by means of filtering and prediction can be incorporated within each sensor, such that if the sensor fails, it gives an indication that it has failed.

Faults in (iii) actuators and (iv) data recorders are generally similar in their end results to sensor failures and usually can be detected by input/output monitoring and /or by the provision of multiple path solution.

The final type of fault to be considered is (v) controller failure, which in the sense of a computer control system generally means an algorithm or a numerical failure.

The control scheme described here can be made to deal directly with all fault types, particularly controller and plant faults. The controller type is considered in a general sense in Figure 5.1, where it is assumed that a computer based control system is employed, such that everything outside the shaded area is, in real terms, computational.

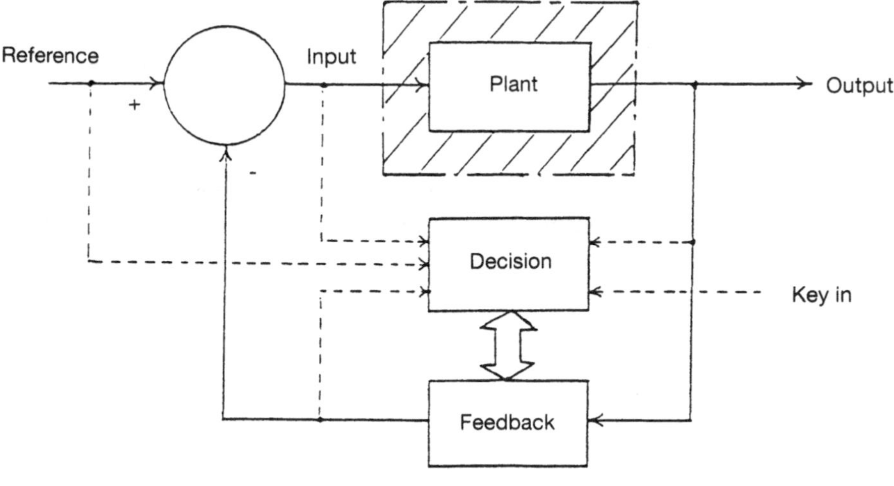

Figure 5.1. General Controller Scheme

The plant output is therefore sensed and fed via an A/D converter into the computer, whilst the output is fed from the computer via a D/A converter to the plant actuator. It is intended that multi-inputs and multi-outputs are allowed for in Figure 5.1, and that the measured output signal can appear via a number of parallel or inter-connected sensors where appropriate.

The extent of the Decision block is considered in Section 5.5, suffice here to say that the dotted lines indicate an information collection exercise. Key in signifying information input from a keyboard, passed from a hierarchical overlord, from interconnected/related systems or from external monitoring equipment. The Feedback block, discussed in Section 5.3, can take on one of a large number of forms, ranging from a simple gain value or digital controller to, as is later described, a parallel arrangement including strategies uch as adaptive, predictive and optimal loops. A simple case of the Decision and Feedback arrangement is a straightforward Gain Scheduling device, operating on feedback gains which are dependent on the output signal value.

To conclude this section it is worth noting that the controller scheme described is to provide a versatile intelligent, plant dependent control device, the complexity of which is dependent on the complexity of the plant, the amount of inter-sample computational time available, computational cost etc. The aim is to make use of the flexibility provided by a computer control base to produce a practical real-world controller, rather than either a mathematical exercise that is of no practical use or a variable controller that cannot be trusted.

5.3 FEEDBACK PATH

In Figure 5.1 it is intended that the feedback allows for a multiple feedback arrangement, taking in one or several output signals. Treating the Decision block as a separate, but related entity, the feedback path can be broken down further as shown in Figure 5.2.

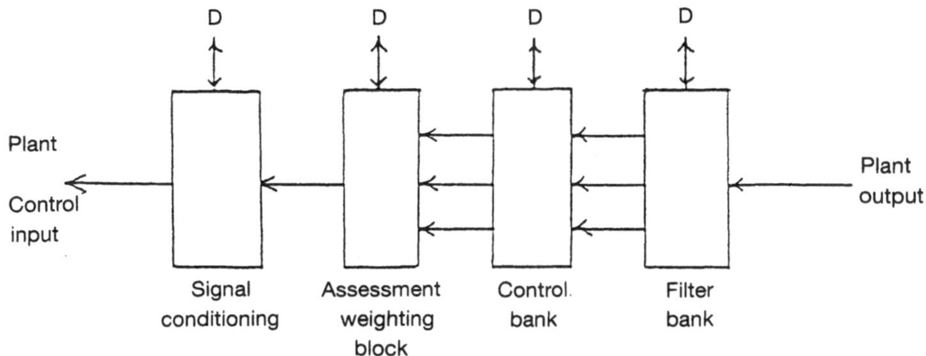

Figure 5.2. Parallel Feedback Path

Although the possibility of multiple output/input signals is allowed for, in order to show how each signal is broken into several parallel loops, only a single input/output signal is employed, so it is considered that a typical single variable feedback loop controller consists of a parallel feedback loop arrangement rather than the usual single loop.

The filter bank is often encountered in practice for such tasks as low pass filtering to eliminate noise on the measured signal. The main difference here is that at this point the original signal is split up into several parallel paths. Therefore if it is desired and/or considered necessary by the Decision block (marked D in Figure 5.2), different filter types can be assigned to different parallel limbs; the filters can though be evaluated concurrently.

The Control Bank is, to all intents and purposes, simply a bank of feedback controllers operating concurrently in parallel, the control action applied in each loop either having been selected *a priori* or, especially where an adaptive form is suitable; having been updated via the Decision Block. It is important to realize that the control block is assumed, in general, to consist of a bank of different controller types, e.g. a P + I controller in parallel with an optimal controller in parallel with a deadbeat controller etc. This then reduces to a single feedback controller as a special case, or as another example a bank of "identical" controllers merely providing a multiple path feedback arrangement for fault tolerance.

The extent of the Decision block action on the controller block is very much dependent on the type of controllers included in the bank. A fixed parameter P + I controller would require little or no intervention whereas a self-tuning control scheme (3) would require periodic updating of its controller parameters. Whatever the type of the controller in any particular loop however, it is expected that a certain amount of controller parameter updating will take place under the influence of the Decision block, as is described in Section 5.5. In order to link the controller block itself directly in with fault tolerant design, one type of control action which is particularly useful is adaptive, predictive control – this can be used to predict output signal values, errors being signalled when the actual measured value differs significantly from that predicted.

The Assessment/Weighting block is described in the following section, where it is shown how the different signals obtained from the parallel controller block, are brought together.

5.4 CONTROLLER ASSESSMENT

As shown in Figure 5.2, the parallel signals obtained from the controller block are fed directly into an Assessment-Weighting block, which is also under the influence of the Decision block.

In simple terms, the Assessment block operates as a summing junction, where different weightings are given to each of the signals arriving from the controller block. The resultant signal leaving the Assessment block is therefore a weighted combination of the various controller forms. As an example, the end result is shown in Figure 5.3.

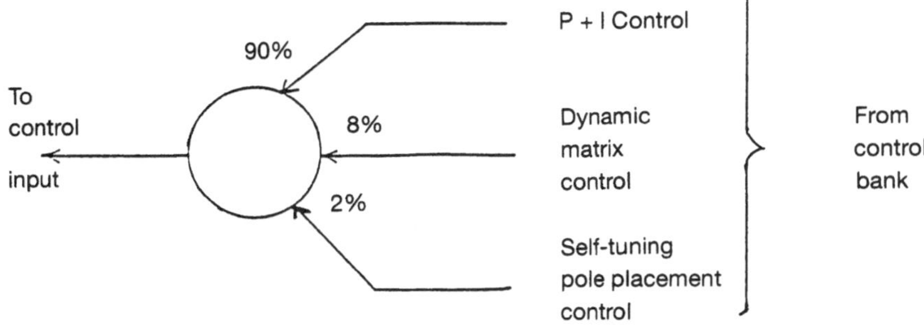

Figure 5.3. Assessment-Weighting Block (an example).

So the signal sent off, via signal conditioning, to form the plant input signal, is made up of several different control signals, each providing a certain percentage of the whole. The percentages/weighting applied to each individual signal is directly selected by the Decision block and depends on the most recently collected information, i.e. operator information via the keyboard or system condition information. Hence control signals can be switched in or out by the Decision block, as required, such that if a controller fault occurs, this limb is simply fed in as 0% of the total. Similarly if it is considered to be highly probable that either a particular controller is faulty or it is causing a problem, then it can be fed in with a very low percentage if this is appropriate.

The Assessment block is also a point at which comparisons can be made between the different control signals arriving at the block, and this can be employed as an error checking-mechanism, i.e. when one control signal is unexpectedly distinctly different from all other control signals then the likelihood of the relevant controller being faulty is much increased. Further, in the simple case when the control block is formed by a bank of identical controllers, the Assessment block can be employed as a majority voting point for example [4].

In order to provide flexibility, the summing weightings/percentages would normally be selected by the Decision block, however it should be noted that provision is allowed for effectively "manual" adjustment by straightforward selection. The usual practice though, would be for the weightings to be time and signal dependent, allowing for controllers to be switched in or out in order to cope with faults should they occur.

The signal conditioning block shown in Figure 5.2 can be employed as a post filter if desired and can be assumed to effect reference input summation if desired, along with feedforward offset allowance. As it is not of major importance in this discussion, it will not be considered further here.

5.5 DECISION AND MONITORING

With reference to Figure 5.1, the Decision block can be extremely complex, the level of complexity depending on (i) the amount of informated collected, (ii) the amount of signal monitoring required, (iii) the number of parallel controller limbs, (iv) the complexity of the individual controllers, and (v) the level of intelligence and reasoning desired. As an example of the last of these, a standard requirement made of the Decision block would be to carry out analyses of signal trends obtained through outcome evaluations based on plant models, in similar fashion to [5].

The Decision block must necessarily carry many error checking routines with action courses set out for each likely error. However the overall block complexity is affected directly by the type of plant under control and the number and type of sensors, specifically as both of these points will affect the type and number of controllers employed.

Where adaptive controllers are employed within the Control block, it is assumed that the Decision block incorporates controller updating procedures, such as those requiring a recursively calculated plant model. Such updating procedures must themselves be assessed, giving an indication of the depth of interrogation considered.

5.6 CONCLUSIONS

The parallel controller structure discussed here is particularly well suited for use on parallel computer architectures, as the concurrent evaluation of different controller forms results in an overall feedback computation time which is equivalent to, at most, that of a single feedback loop arrangement.

The scheme has the following immediate advantages :

- It allows for a conservative use of modern, possibly less robust, controllers which provide a fine tuning action for performance improvement.
- It allows for one (or a number of) controller(s) to be switched in, whilst a faulty controller is taken off-line, or whilst a plant/sensor fault is corrected.
- It allows for performance assessment and confidence building with regard to a new controller – without risking plant shutdown.
- The total cost is low, especially if only computer control is employed, such that only software modifications are required in order to bring new controllers on-line.

REFERENCES

1 Fleming, P.J. (Ed), "Parallel Processing in Control", Peter Peregrinus Ltd., 1988.

2 Warwick, K., "Control Systems : An Introduction" , Prentice-Hall, 1988.

3 Warwick, K. (Ed), "Implementation of Self-tuning Controllers", Peter Peregrinus Ltd., 1988.

4 Ham, P.A.L., "Reliability and Redundancy in Microprocessor Controllers", Chapter 12 in "Industrial Digital Control System", K. Warwick and D. Rees (Eds), Peter Peregrinus Ltd., 2nd edition, 1988.

5 Ashmole, P.H., "Power System Alarm Analysis and Fault Diagnosis using Expert System", Chapter 15 in "Failsafe Control Systems", K. Warwick and M.T. Tham (Eds), Chapman and Hall, 1990.

Chapter 6

Fault Tolerant Procedures For Boiler Control Systems Design

R. Clarke
CEGB

6.1 INTRODUCTION

The use of digital control systems within the CEGB has grown over the last decade and today there are approximately seven hundred digital computers performing control and monitoring functions on power stations throughout the country [1]. The prime consideration in the design of such systems is to ensure that a fault in any computer or its associated peripherals does not jeopardize the integrity of the plant which is being controlled/monitored. An accepted standard of "good practice" has been developed for the design of process control/monitoring systems and this can be illustrated by reference to a major refurbishment recently undertaken at Didcot Power Station.

6.2 MOTIVATION AND DESIGN CRITERIA

Didcot Power Station is situated on the Thames near Oxford and operates four coal fired boiler/turbine units, with a generation capacity of five hundred megawatts per unit.

The station is required to adjust its generated output continually in response to the demands of the National Grid. This often entails shutting down units at night and restarting them the following morning, ready for the peak demand at breakfast time. Such an operating regime imposes a heavy burden on the plant and its associated control/monitoring system. In order to satisfy the increasing requirements of the National Grid without imposing undue stress or damage upon the plant, a major refurbishment of the Station's control/monitoring systems was undertaken.

The benefits gained from the refurbishment include :

- Reduced startup/shutdown times;
- Improved plant efficiency;
- Improved reliability/maintenance;
- Improved operator interface.

However, all these benefits would be outweighed by the cost of a major fault on a boiler/turbine unit. It is thus imperative that the control and monitoring system should "fail safe". For some applications, such as a nuclear power plant or aircraft, it may be necessary to provide triple channel redundancy or alternatively a main and standby system in order to maintain a fully active system in spite of equipment failure. On conventional power plant, however, the operator is able to intervene provided that the fault has not caused the simultaneous loss of too many functions, or taken the plant into an unsafe operation region. This imposes the design requirements of "graceful degradation" and "freeze fail".

6.3 SCOPE OF THE PROBLEM

Figure 6.1 shows a schematic diagram of a coal fired boiler and illustrates the major control functions.

Forced Draught (FD) Fans : Two variable throughput fans used to force air into the furnace; safe and efficient combustion depends upon their correct operation.

Induced Draught (ID) Fans : These two fans are larger than the FD fans and extract the combustion products from the furnace, which must be maintained at 0.25 bar below atmospheric pressure.

Feedwater Control : The water level in the boiler drum is controlled by a steam turbine feed pump and two out of three feedwater valves.

Attemperator Spray Flow : Water is sprayed into the steam immediately before the final superheater stage. This controls the temperature of the steam passing from the boiler into the turbine.

Pulverizing Fuel Mills : These grind the coal into a fine powder and blow it into the furnace. Each boiler/turbine unit has eight mills and requires six mills to be in service in order to generate full output.

Boiler Pressure/Unit Load Control : The total firing level in the furnace and the steam flow to the turbine are controlled in order to maintain the required boiler pressure and unit load.

In total there are 85 modulating loops per unit together with numerous isolating valves and dampers. In addition the scheme includes diverse monitoring functions such as: turbine vibrations, boiler metal temperatures, generator temperatures and feedwater chemistry. Altogether, the scheme monitors/controls over 1,000 analogue inputs and 3,000 digital signals per unit.

6.4 COMPUTER ARCHITECTURE

In order to limit the effect of a computer failure and to achieve graceful degradation, the distributed architecture shown in Figure 6.2 was adopted. This consists of 20 DEC11/73 computer systems linked by a dual Ethernet communications highway.

The control and monitoring functions are distributed in such a way that if any computer fails the operator can continue to run the plant safely. This requires the "double scanning" of a limited number of signals and the provision of a "backup" panel with hardwired control switches and conventional instruments.

The control functions have been allocated on a plant area basis; but with a view to achieving redundancy through diversity. For example, only one FD and one ID fan control functions are allocated to a single processor, so that in the event of failure, furnace conditions can be controlled by the other pair of fans. Ten of the processors shown in the lower half of Figure 6.2 have been allocated to control functions, and a further three processors are used to perform various monitoring functions. The final processor in the lower group is designated the system host and performs a housekeeping/system management role.

The remaining six processors form the basis of a "soft desk", and each has a 9 inch monochrome touch pad and a 14 inch touch screen with colour display connected to it. The soft desk provides the operator with the facilities to :

- Initiate and monitor plant startup/shutdown sequences;
- Open/close isolating valves/dampers;
- Monitor plant variables and display "trends";
- Acknowledge and accept alarms;
- Select loops to auto/manual;
- Enter setpoints;
- Obtain diagnostic information.

All the data and commands are transmitted over the dual Ethernet and any "desk module" can carry out any function; thus providing a highly fault tolerant interface to the operator.

Two potential sources of common mode failure are the power supply to computer suite and data highway. The use of an uninterruptable power supply guards against the former and the use of dual Ethernet provides a redundant data highway. Experience to date indicates that only a small number of node failures have occurred and these have not affected the integrity of the remaining nodes.

The chosen architecture has satisfied the criteria of "graceful degradation" and provides the additional benefits of :

Flexibility : Display formats can be modified and new ones added, to meet changing operational requirements.

Expandability : additional computer can be linked in to carry out new functions, e.g. flue gas desulphurization.

Incremental Commissioning : The refurbishment of control and monitoring systems is carried out during plant "outage" periods. It would have been extremely difficult to install a scheme of this size using a monolithic computer structure.

Modularity : The repeated use of common software and hardware components leads to reduced maintenance and spares holdings.

6.5 CONTROL LOOP STRUCTURE

In order to achieve the second design criterion of "freeze fail", the control loops and their interface to the plant must be designed and structured correctly. Figure 6.3 is a detailed section taken from Figure 6.2 and identifies the major elements of a digital control loop namely :

- Input of plant data to the computer;
- Output of commands to plant actuators;
- Implementation of the control algorithm;
- Operator interface.

One feature of this design is that plant signals are not connected directly into the main computer; but enter via serial lines from intelligent "field units" such as scanner or PLCs. Fibre optics was the preferred technology for the serial lines, but it was ruled out on cost. The alternative was to adopt the RS 422 standard and to install data line protectors to guard against possible voltage transients. There is always a large amount of "electrical noise" present on power stations and the use of the serial lines effectively prevents this from corrupting the operation of the computer systems. The serial lines also provide a convenient interface point for installation, commisssioning and ultimate trouble shooting on the system.

In addition to the components shown in Figure 6.3, the operator is provided with "backup" panel from which he can manually operate items of plant in the event of a computer system failure.

Each element of the control loop can now be examined and potential faults identified.

6.6 SCANNERS AND DATA INPUT

All of the analogue signals and the majority of the digital signals associated with control loops enter the system via digital scanner units. The analogue channels are organized into groups of 16 with a 12 bit A/D converter and multiplexer per group. This allows a "mains" integrating A/D converter to be used and still achieve a 1 sec scan rate. In addition to mains rejection capabilities, the scanners have a common mode rejection ratio of 50V and are able to withstand a possible 2 kV earth fault which could be induced by an electrical fault on a generator.

The scanners have a built-in microprocessor and perform self-diagnostics. Further integrity checks can be performed by monitoring "check" channels on each scanner. "Active" check channels have been used on some previous projects but it is now felt that the additional security does not justify the increased complexity.

The plant signals connected to the scanners are designed to be 4-20 mA wherever possible. In addition to checking for the "active zero" the computer system converts each signal into engineering units and checks it against physical limits.

The physical limits include maximum and minimum values and also a maximum rate of change check. In certain cases there is some redundancy and it is possible to cross-check groups of signals; for example valve position, differential pressure and flow should match known characteristic relationships. Work is currently being carried out in the universities to provide "dynamic" fault checks based on Kalman filtering techniques, these approaches are being examined, but have not been used operationally on the present project [2], [3].

The failure of any integrity check will render the associated signals invalid prevent any control loop which uses them from being selected to auto control.

6.7 OUTPUT COMMANDS AND ACTUATOR DRIVES

All the modulating control loops are driven through PLCs via RS 422 serial lines. Many of the major plant items have startup/shutdown sequences associated with them. The appropriate PLC will perform these sequences on command from the main computer system and once the plant item is in service it will increment or decrement the actuator position in response to commands sent down the serial line. The PLC has a separate stepper drive module connected to it for each loop under its control. This interfaces to either a pulse to pneumatic or pulse to current converter containing a four phase stepper motor. The output "chain" thus depends upon incremental signals and will naturally "freeze" in the event of a computer system failure.

Additional output checks are provided by monitoring the position of the stepper motor and final actuator. Discrepancies between these measurements and the computed actuator position can indicate faults in the plant wiring or actuator linkages.

In the event of an "output" fault being detected, the associated control loop cannot be selected to auto. However, the operator may be able to drive the actuator manually from the "backup" panel are wired directly into the stepper drive modules and will operate even if the computer and PLC have both failed.

6.8 CONTROL ALGORITHM

The incremental nature of the computer output commands leads to a natural choice of an incremental control algorithm for all the modulating control loops. The algorithm takes the general form :

$$\Delta U_k = a_1 \Delta U_{k-1} + a_2 U_{k-2} \ldots a_m U_{k-m}$$

$$+ b_0 e_k + b_1 e_{k-1} + b_2 e_{k-2} \ldots b_m e_{k-n}$$

$$+ c_0 x_k + c_1 x_{k-1} + c_2 x_{k-2} \ldots c_p x_{k-p}$$

where ΔU_k = Required incremental change in the actuator position

e_k = Error between measured value and setpoint

x_k = Optional feed forward or cross-coupling variable.

The number of coefficients a_i, b_i and c_i together with their values are determined as part of the design procedure for each loop. The design is normally carried out with the aid of a commercially available CAD program; but for some of the simpler loops the general form degenerates into an incremental PID structure and the coefficients can be obtained by "digitizing" an analogue design. However, where there is interaction between associated loops or significant time delays the general structure gives a better response and is preferred.

The incremental nature of the algorithm provides a natural "freeze fail" mechanism. It also provides a simple form of bumpless transfer when changing from manual to auto control and does not suffer the integral wind-up problems of conventional designs.

6.9 OPERATOR INTERFACE

Inherent in the system design is the assumption that the operator is able to intervene in fault situations. As a last resort he is able to use the "backup panel", but because of the built-in diversity and redundancy of the "soft desk" he will still have most facilities available to him even under severe fault conditions.

In these situations, an important feature is the ability to recognize that a fault has occurred and to diagnose its cause. It has been shown that various input/output faults can prevent loops from operating in auto mode. When such a fault occurs, the software will trip the loop to manual control and sound an audible alarm.

The displays on the 14 inch touch screen will indicate where the fault has occurred by flashing an appropriate indication in red. The operator can touch the flashing legend to call up a subsidiary display showing the control loop which has failed. From this display he can select a diagnostic display on the 9 inch touch pad, which will indicate the cause of the fault. Resources can thus be employed more effectively in the rectification of faults and maintenance of the system.

6.10 SOFTWARE STRUCTURE

In addition to input/output faults, control loops can be tripped from auto control by system/software faults. These may be due to incipient programming errors which were not detected by the commissioning procedures or they may be caused by a hardware fault in a computer's CPU or memory card. Parity memory is now widely available and will detect memory errors, but further safeguards are required to ensure the integrity of the software.

The system software used was developed in-house and provides a real time, multi-tasking, multi-processing, "high level" environment in which the engineer can write applications programs. Central to the software is the concept of a job slot. This is a block of engineering program which can occupy up to 4K words of memory and can be set to run just once or at a fixed interval, eg every second. Alternatively a job may be configured to run to completion and then "sleep" either for a fixed period of time or until it is "reactivated" by a command from another job.

The software also includes the concept of "global" data. This allows information to be passed from one job to another and also from one processor to another via the dual Ethernet communications highway. The system software has built-in error checking and facilities to enable the engineer to include "error recovery" procedures in his code.

A typical control processor would include jobs of the following form :

Scanner Job : This interfaces with the scanner unit, converts the data to engineering units and validates it.

Setpoint Job : Handles requests for setpoint changes entered by the operator, provides bumpless transfer on A/M changeover.

Interlocks Job : Trips loops from auto if fault is detected.

Alarms Job : Checks data against alarm levels and also alerts operator if a loop trips.

Trend Job : Stores past values of every analogue input on Winchester disk and makes data available to the operator on request.

Control Job : Implements the control algorithm and calculates the required actuator increments.

Actuator Job : Interfaces with the PLC/stepper drive module and checks for output faults.

Monitor Job : Software Watchdog facility.

The function of the last job is to check that every other job in the processor is operating as scheduled. This is achieved by "global" counters which are set in each job when it has run to completion. The monitor job runs at the top priority and decrements all the job counters every time it runs. If a counter ever reaches zero it implies that a fault has occurred to prevent the associated job from running. Control loops can be tripped as necessary and an alarm sent to the soft desk. In order to protect against failure of the monitor job or indeed a complete processor, a "master" job in the host processor monitors the monitor jobs in all the other processors. The host is unique in possessing a hardware "watchdog" card containing a relay which must be "held in" by continually addressing it from the processor. Failure to "hold the relay in" will initiate a hard wired alarm indicating failure of the system host, in this situation the rest of the system will function as normal but without the master watchdog facility.

6.11 CONCLUSIONS

This chapter has outlined the fault tolerant concepts incorporated into the refurbishment of the control and monitoring system of a conventional power station. Fault tolerance must be considered at every stage of the design process and cannot be "added" at the end.

In the project considered, "fault tolerance" was based upon the two criteria of "graceful degradation" and "freeze fail". These were achieved by adopting a distributed computer architecture and by correct design of incremental control loops.

REFERENCES

1 Waddington, J. and Maples, G. C., "Control of Large Coal and Oil Fired Generating Units", Power Engineering Journal, No. 1, January 1987.

2 Eckert, S. J., Loparo, K. A. and Roth, Z. S., "An Application of Non-Linear Filtering to Instrument Failure Detection in a Pressurised Water Reactor", Nuclear Technology, Vol. 74, August 1986.

3 Wallace, J. N. and Clarke, R., "The Application of Kalman Filtering Estimation Techniques in Power Station Control Systems", Trans. IEEE, Vol. AC-28, March 1983.

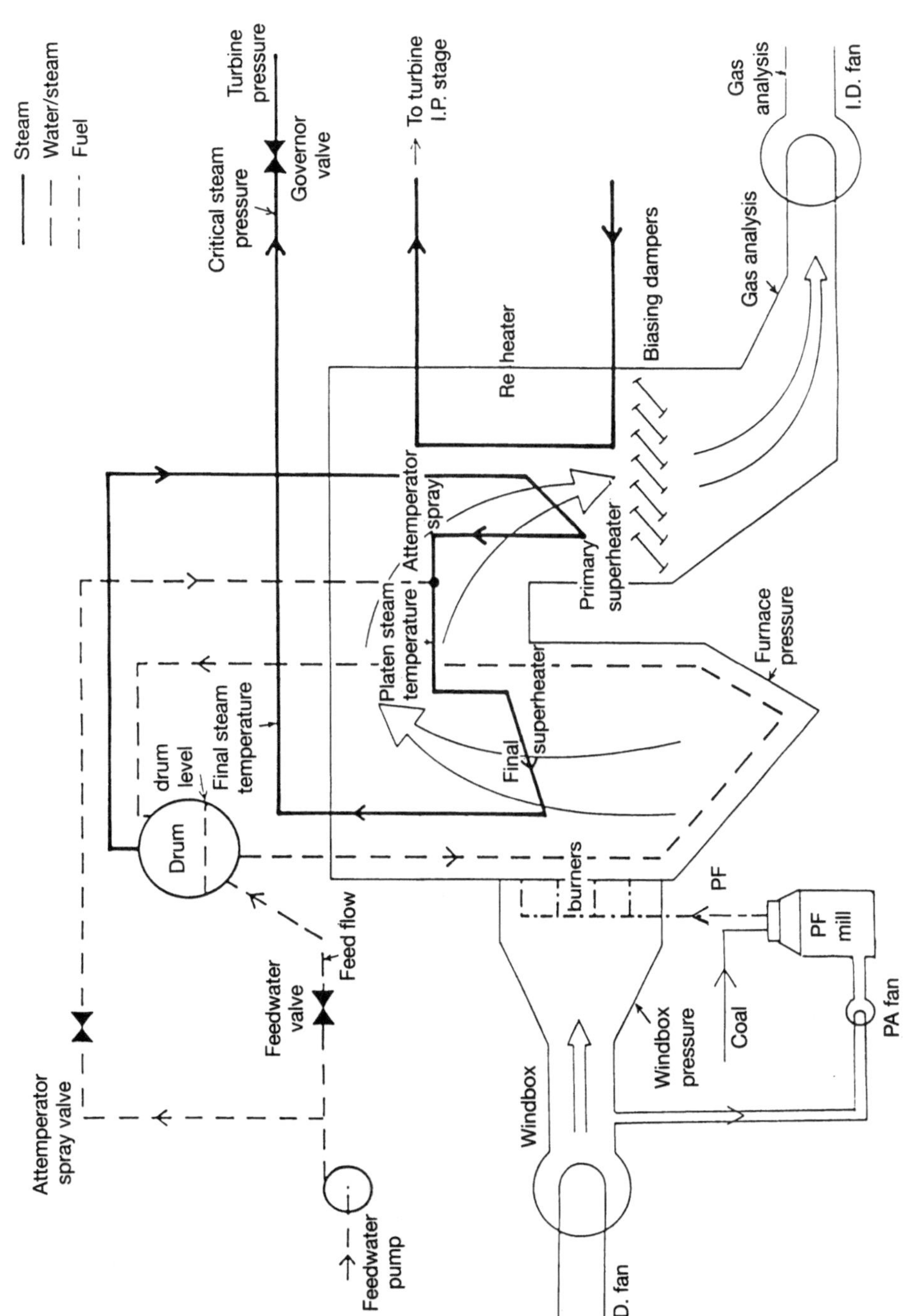

Figure 6.1. Schematic diagram of a coal fired boiler.

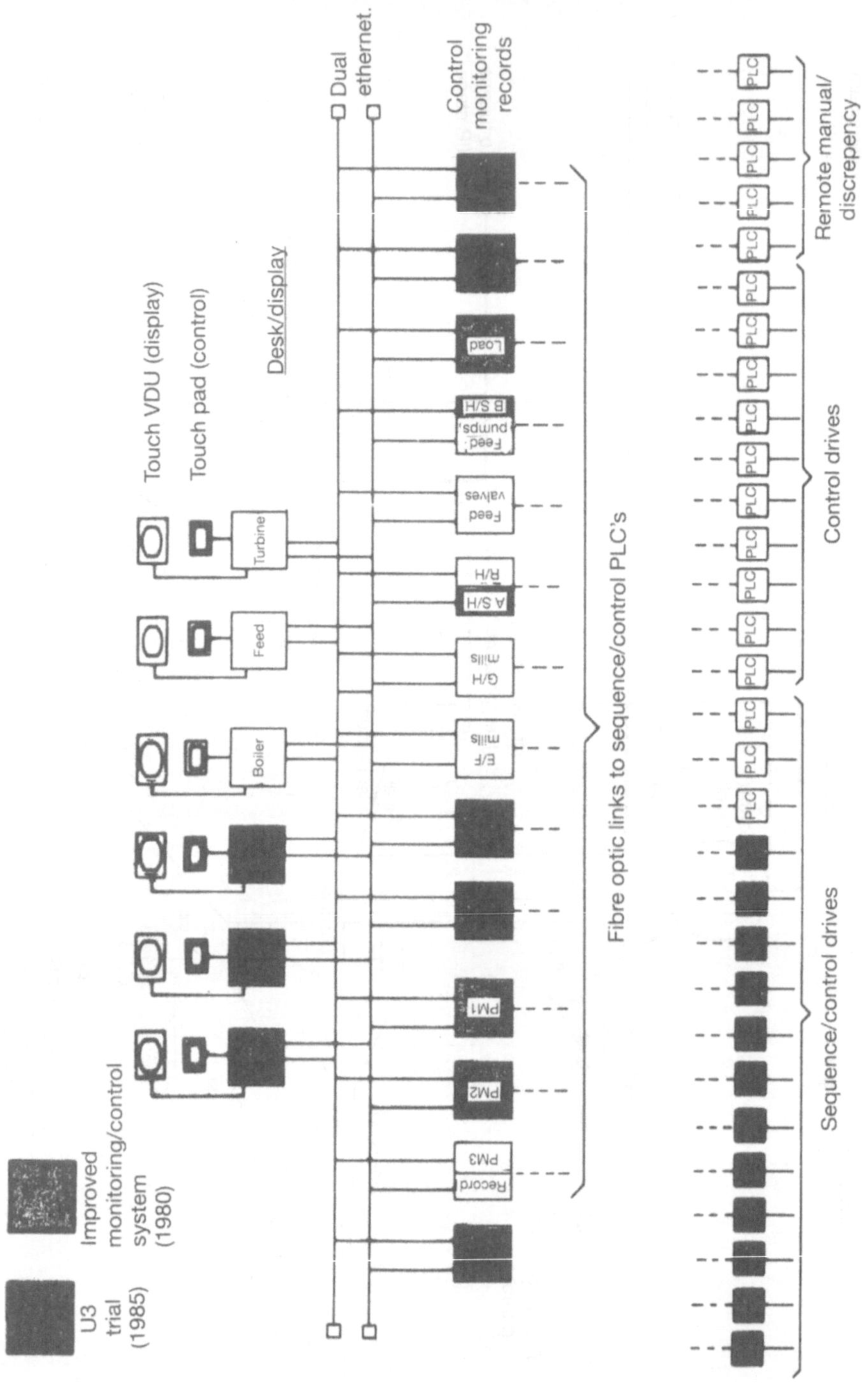

Figure 6.2. DIDCOT C & I scheme – Computer structure.

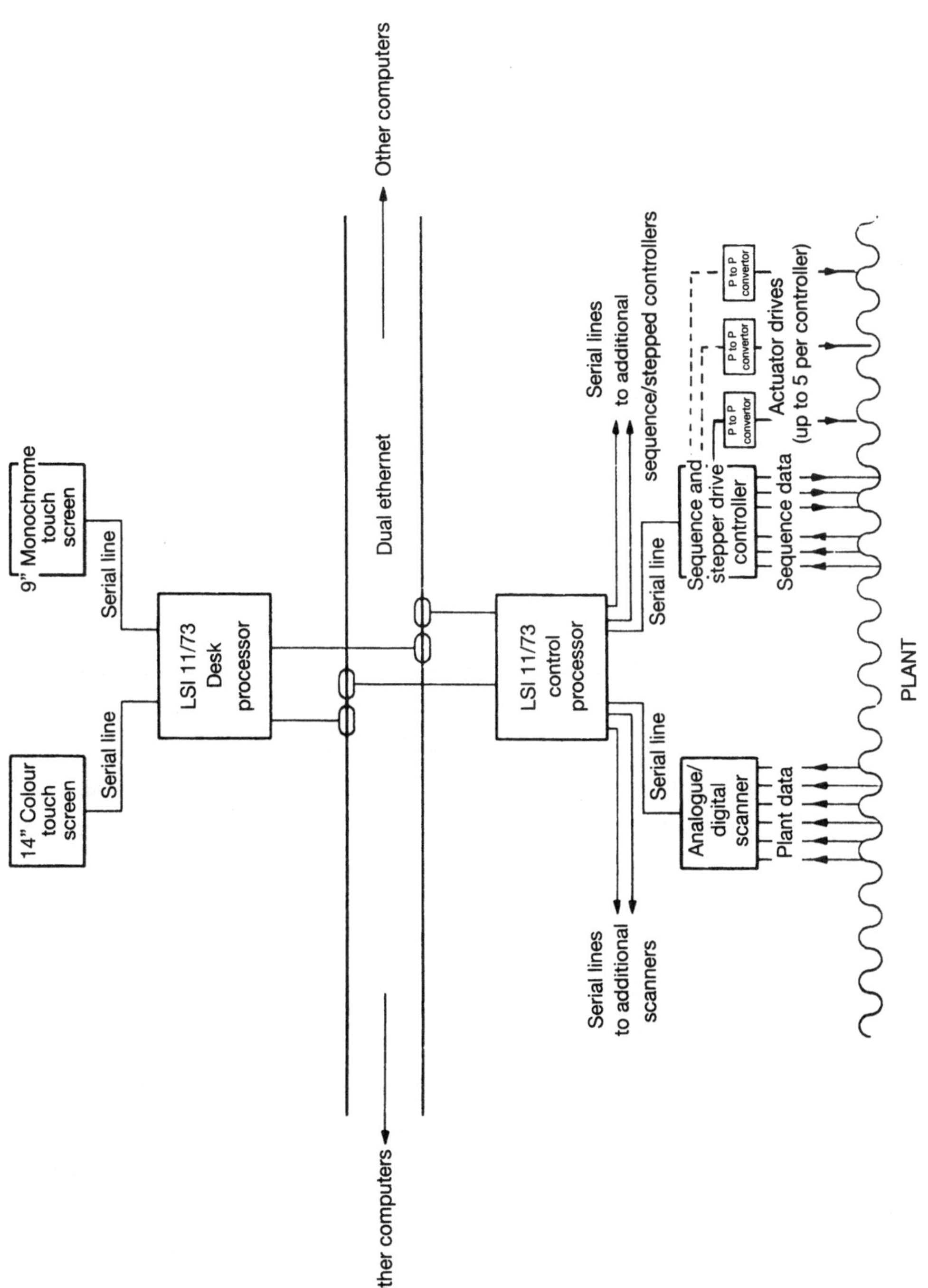

Figure 6.3. Typical control structure.

Chapter 7

FRAMEWORK FOR THE DESIGN AND ASSESSMENT OF SAFETY RELATED CONTROL SYSTEMS

R. Bell; M.F. Pantony
Health and Safety Executive,

7.1 INTRODUCTION

Computer based systems, generically referred to as programmable electronic systems (PESs) have been used in the process industries for many years – particularly for process control functions. Both centralized and distributed systems have been extensively used. There is no doubt that this trend will continue, and in fact accelerate, due to the many advantages such systems offer the plant operator. The realization of the advantages will, however, only come about if a disciplined and structured approach to the design is adopted at all project stages. A timely reminder of this is illustrated by a recent failure of a computer system controlling a nylon polymer plant which led to a serious plant incident [1]. It has been estimated that the cost of the post-incident assessment was ten times what it would have been had a proper assessment been carried out at the project design stages.

In the past the role that PESs have played have been largely restricted to *process control functions* and have only played a secondary safety role. However, the operational and cost advantages of using PESs are now being exploited in the context of protection systems having a primary safety role. *There is thus an increasing trend to provide both process control functions and protection functions by means of computer systems*. Such systems not only offer the potential to achieve higher levels of safety integrity but also offer the plant user advantages by way of reduced operating and maintenance costs together with ability to perform complex interlocking and plant monitoring functions. If UK industry is to maintain its competetiveness it is important that the potential financial benefits are fully realized. *Yet this must be done whilst achieving an adequate level of safety.* The potential for improved levels of safety integrity is significant. However, the level of complexity involved means that improvements will only come about if a throughly considered design and assessment methodology is adopted. In order to provide such a methodology, the Health and Safety Executive (HSE) published two documents in June 1987, which are the first in a series whose general title is "Programmable electronic systems in safety related applications". The two documents are:-

(1) "**An introductory guide**"[2]. This document ("**PSE 1**") is aimed at the non-specialist and provides an overview of the safety principles.

(2) "**General technical guidelines**"[3]. This three part document ("**PSE 2**") contains:

- general guidance on the problems, and a framework within which they can be approached systematically (Part 1).
- a method for assessing the safety integrity of PSEs – including the hardware and software. (Outlined in Part 1 and described in detail in Part 2).
- a worked example using the method in Part 2 is described in Part 3.

The guidelines are *generically* based and should enable the safety integrity of systems incorporating PESs to be determined irrespective of the application. They have been structured so they do not unreasonably constrain design innovation but allow programmable electronics technology to be safely exploited. A major objective in producing such generically based guidelines was to stimulate industry, and others, to produce their own guidance for specific applications. It is HSE's policy to encourage and give help in the development of this *application specific guidance.*

This chapter:

- provides information on guidelines developments taking place within HSE, industry and national/international standards bodies.
- provides a brief overview of the guidelines. Other recent papers have examined them in more detail or have considered particular facets of them [4, 5, 6].

- considers, in particular, the application of the guidelines in the context of a process environment — in particular those situations where the control functions and the protection functions and the protection functions are wholly or partly dependent on PESs.

7.2 SYSTEMS UNDER CONSIDERATION

The guidelines are concerned with those PESs which either acting alone or in combination with non-programmable systems, provide the required level of safety. Such systems, upon which the safety integrity of the plant relies, are referred to as *safety related systems.*

The guidelines do not apply if an *adequate* level of safety is assured by one or more separate non-programmable systems of *conventional safety integrity** or better. Such conventional systems will need to cater for, amongst other things, failures of the controlling PES.

***Note:** The term *conventional safety integrity* means the level of safety integrity which has been achieved, in similar situations, by conventional safety related systems which have traditionally been accepted as good engineering practice.

The PES is defined as a system based on a computer connected to sensors and/or actuators on a plant for the purpose of control, protection or monitoring. The term includes all elements in the system extending from plant sensors or other input devices, via data highways or other communication paths, to the plant actuator, or other output devices.

The part of the PES which handles the logic processing is termed the "**programmable electronics**" and refers to those parts of the PES which are not solely dedicated to a particular sensor or actuator on the plant and in which different functions are performed at different times under the control of software. *The term therefore includes both software and hardware elements*. Figure 7.1 illustrates the basic PES structure. In the context of a safety related system using a Programmable Controller (PC), the programmable electronics would reside within the PC.

7.3 CONSIDERATIONS UNDERLYING THE GUIDELINES

To ensure safe operation of safety related PESs, it is necessary to recognize the various possible causes of PES failure and to ensure that adequate precautions are taken against each. Two basic types of failure are considered – **Random hardware failures** and **Systematic failures.**

Random hardware failures are those failures which result from a variety of normal degradation mechanisms in the hardware. Measures of reliability such as the "mean time between failures" (MTBF) are concerned only with random hardware failures and do not include systematic failures.

Systematic failures are concerned with errors in the design, construction or use of a system which cause it to fail under some particular combination of inputs of under some particular environmental condition. Failures arising from incorrect specification, errors in the software and electrical interference are all examples of **systematic failures.**

7.4 SAFETY PRINCIPLES

The safety strategy underlying the recommendations made in the guidelines are centred on three system characteristics or *system elements*. The principles which govern the *system elements* underlie the design and assessment strategy for a safety related PES. The three system elements are defined as follows:

- **Configuration**: The specific arrangement of the programmable electronics within a PES and the combination of PES and non-PES safety related systems.
- **Reliability**: That aspect of the safety integrity relating to random hardware failures in a dangerous mode of failure of the safety related systems.
- **Overall Quality**: The non-quantifiable qualitative aspects of the safety integrity of the safety related systems. This system element is concerned with the precautions taken against systematic failures.

The detailed requirements of the three system elements are, together, intended to tackle both random hardware failures and systematic failures (Figure 7.2). The safety integrity level for the safety related systems is specified in terms of the three systems elements – the exact package of which will depend upon the application in question and therefore the level of safety to be achieved. This package constitutes the *safety integrity criteria* for the application. For a specific application the three system elements will be specified as follows:

- The **configuration** will be specified in terms of the number of safety related systems together with the requirements relating to the programmable electronics (both hardware and software).
- The **reliability** will be specified either *qualitatively* or *quantitatively*.
- The **overall quality** will be specified in terms of the precautions that need to be built into the design, operation, use etc, against systematic failure causes. For those quidelines use a series of checklists which are organized such that each checklist relates to one of the 16 life-cycle phases (Figure 7.3). *The purpose of the checklists is to provide a stimulus to critical appraisal of all aspects of the safety related systems rather than lay down specific requirements.*

The safety principles relate to the *total configuration of safety related systems* required to achieve an adequate level of safety integrity for the hazard in question. The total configuration will comprise, in many cases, both PES and non-PES safety related systems – which may be automatic or manual in operation.

It is recognized that an adequate level of safety integrity may be achieved other than by the strategy put forward in the HSE guidelines. HSE believe, however, that the strategy recommended represents a practical foundation on which to base the design, taking into account all potential causes of failure including software faults and electromagnetic interference.

7.5 DESIGN AND ASSESSMENT GENERAL FRAMEWORK

The overall framework, including the key steps, for design and assessment of safety related PESs is shown in Figure 7.4.

From Figure 7.4, it can be seen that:

- the required level of safety integrity is specified in terms of the three system elements – *configuration, reliability and overall quality. This specification for the safety integrity in terms of the system elements constitutes the safety integrity criteria*
- the *safety integrity criteria* relates to the total configuration of safety related systems (both PES and non-PES)
- the *safety integrity criteria* are used as the basis of design and analysis of the safety related systems.

It is intended that future guidance documents will specify the *safety integrity criteria* for specific applications. Where no such criteria has been developed for the particular application, the overall objective should be to ensure that the safety integrity of the total configuration of PES and non-PES safety related systems should not be inferior to *conventional safety integrity*.

In determining *conventional safety integrity* for a particular application, guidance may be obtained from consideration of conventional systems which have been or would be accepted in similar circumstances. In some cases, it will be possible to make a direct comparison with existing or replaced plant. A direct comparison will not always be easy since PESs are used in many new fields of technology and applied in many new ways; in such cases, accepted good practice in other similar situations will provide important guidance.

The design and assessment framework (Figure 7.4) has been discussed in terms of the total configuration of safety related systems one or more of which was a PES. However, the framework and much of the guidance contained in "PES 1" and "PES 2" is applicable to non-PES systems. *For example, in the context of Steps 1-6; all steps are relevant to situations where there are no PESs in the total configuration of safety related systems.*

7.6 APPLICATION OF THE SAFETY PRINCIPLES

Publication "PES 2" provides a number of examples of how the safety principles apply in various general cases for those situations in which no safety integrity criteria, specified in terms of the system elements, have been developed. Examples are given for *protection systems; separate control and protection systems; combined control and protection systems; and continuous control for safety*. The examples are not intended to cover all situations or all means by which the safety principles may be satisfied. They are intended merely to illustrate how the safety principles apply in practice. It is intended that there will be further documents to show how they apply in specific circumstances.

7.7 TOTAL SYSTEM ENVIRONMENT

The application of the PES guidelines to the process industry needs to be considered in the context of the "**total system environment**" as it relates to the hazard(s) in question. All those systems which play a part, to a greater or lesser degree, in preventing the hazard(s), or mitigating the consequence of the hazardous event(s) need to be considered. The *total configuration of safety related systems* is a part of this "**total system environment**".

In the context of a chemical plant the key features of this "**total system environment**" are shown in Figure 7.5 and include:

- The main process control system which is designed to keep the plant within its designed operational envelope.
- The operator and his role in the overall scheme of things.
- The alarm system and its role in the overall scheme of things.
- Those systems that have primarily been designed to provide the requisite level of safety (i.e. deviated protection systems).

Only when the above features have been identified and the design philosophy worked out is it possible to make soundly based judgements about a number of issues as they effect the overall safety integrity. For example, the importance of the following:

- The control system contribution to the overall safety. For example, what is the role of the control system in achieving the required level of safety? If other systems are primarily responsible for safety, has a failure rate been ascribed to the control system? How was it formulated?

- Functional specification of the protection systems. How was this formulated? Does it take into account the total configuration of safety related systems? Does it consider all reasonably forseeable events with regard to control system and plant failure causes?

- The demand rate used as a basis for the protection system design. How was this demand rate formulated?

- The role of the operator. Has the operator's performance been taken into account in any estimated control system failure rate?

- Alarm management systems. How have these been taken into account in estimating the control system failure rate? Where they taken into account in estimating the demand rate on the protection systems?

- Software change procedures. The highest degree of formality should be applied to the safety related systems. It may be possible to relax the procedures for other systems – but only after consideration of their function in the overall scheme. For example, for non-safety related systems (as per definition in "PES 2") it should be possible to have less rigorous software change procedures. This is one of the benefits of using the concept of the *safety related system*.

- Maintenance priorities. How have they been determined between those systems which are safety related and others?

The adoption of a structured design and assessment framework should enable the above questions, and other to be answered and enable decisions which affect safety, and which could have important economic implications, to be made on a rational basis. It is important that the "**total system environment**" is developed in the future so that individual elements in this "*environment*" can be considered together with the interaction between each element.

7.8 PROTECTION SYSTEMS

(a) Terminology

The two terms – **SAFETY-RELATED SYSTEM** and **TOTAL CONFIGURATION OF SAFETY-RELATED SYSTEMS** – are fundamental to the guidelines and are considered below in the context of *separate control and protection systems.*

Consider a plant which is controlled by a *main process controller* and *two separate protection systems*. The main controller provides the full range of control functions for the plant but should this controller fail in some way, or conditions on the plant deviate to the extent that it cannot be controlled by the controller, then protection is provided by the *two separate protection systems*.

Failure of the main process controller to keep the plant within its operational envelope puts **DEMANDS** on the two protection systems provide an adequate level of safety, taking into account the hazard in question and the number of **DEMANDS** made upon the protection systems. Each protection system is a **SAFETY-RELATED SYSTEM** and the two protection systems together constitute the **TOTAL CONFIGURATION OF SAFETY-RELATED SYSTEMS.**

In many cases the *main process controller* will have safety functions, what is of importance is that the two protection systems provide, in their own right, the requisite level of safety. *The very fact that the main process controller has safety functions does not of itself make it a safety-related system.*

The concept of the **SAFETY RELATED SYSTEM** has been developed to separate the complexity of the main process control computer from the dedicated protection systems. This has important economic and safety benefits. It is essential however in doing this that an appropriate demand rate is used as a basis for the protection systems. Control system failure will be only one source of demands for protection and overall demand rate on the protection systems may not be sensitive to control system failure. It may not therefore be necessary to carry out a quantitative assessment of the control system in order to determine its failure rate.

The demand rate is used in the design of the protection systems, in the first instance, to meet the requirements of the **reliability** system element. The requirements relating to **configuration and overall quantity** are applicable only to the protection systems and not to the main process controller (providing the main process controller does not constitute a safety related system). This has important advantages with regard to maintaining the initial design safety integrity.

(b) Process plant example

Considering the process shown in Figure 7.6 the vessel contains a liquid in which there would be serious safety implications if the level rose beyond a critical point. The basic system is as follows:

(1) On the lower part of the vessel is a level transmitter feeding into the process computer. The signal from the level transmitter provides:

- Control of the valve which itself controls the liquid in the vessel
- Level indication for the process operator
- An alarm set at a prescribed value, e.g. 80% level. This alarm operates through the process computer software (i.e. it is "software-based").

(2) Above the level transmitter is a single level switch which is hardwired into an alarm.

(3) Above the single level switch are three level switches feeding into two programmable controllers (PC 1 and PC 2). Within each PC "2 out of 3" voting takes place. The output of each PC goes to the trip valve and also to a trip alarm.

(4) A trip valve can also be operated by direct means from a stop button which is hard-wired.

(c) Determination of safety related systems

In the context of the process plant indicated in Figure 7.6; the way in which the **SAFETY-RELATED SYSTEMS** are determined is described below (Figure 7.7).

From a *control and protection* viewpoint, there are essentially four systems.

(1) As the level of liquid in the vessel rises, (to, say, the 80% level), Alarm 1 is raised. The plant has been so designed that the operator can, on receipt of *Alarm 1*, take corrective action to prevent the liquid level from rising any further and, in fact, to reduce the level. *This is system 1.*

(2) If, however, the operator cannot control the liquid level or, the process computer is incapable of taking the corrective action, even though the operator has performed correctly, the liquid level may rise until the hardwired *Alarm 2* is raised. Even at this point, the operator should still be able to take corrective action to stop the liquid level rising any further. *This is system 2.*

(3) If the operator cannot control the liquid level and the liquid continues rising, it will then activate, via the three level switches and PC 1 and PC 2, the trip valve to automatically bring the plant to a safe state. *This is system 3.*

(4) Should the system fail, then the operator can bring the plant to a safe state by operating a hardwired system activated by a shut-down button. *This is system 4.*

In this particular case, the *safety-related systems are systems 3 and 4*. These two systems provide the requisite level of safety, taking into account factors such as the level of hazard involved and the number of demands made upon *systems 3 and 4 by failure of systems 1 and 2.*

It can be seen that whether a system is **SAFETY-RELATED** is determined by the particular circumstances and many factors need to be considered. The role of the operator in *systems 1 and 2* is very important, since efficient and correct operator action will minimize the **DEMANDS** on the two protection systems (*systems 3 and 4*). This has an important bearing on the level of integrity required of *systems 3 and 4* in order to achieve an acceptable **HAZARD RATE.**

As indicated previously, the guidelines apply to the **TOTAL CONFIGURATION OF SAFETY RELATED SYSTEMS** – in this case *systems 3 and 4.*

The identification of the *safety related systems* is Step 2 in the design and assessment framework (Figure 7.4). Steps 4-6 then need to be carried out. Step 3, the determination of the required level of safety integrity, is a vital step prior to any assessment. The HSE guidelines provide guidance on how this can be obtained if no established safety integrity criteria *for the application* exists. The process plant control and protection systems indicated in Figures 7.6 and 7.7 is an example of *separate control and protection systems.*

7.9 SAFETY CASES FOR CIMAH

The **Control of Industrial Major Accidents Hazards (CIMAH) Regulations** requires under Regulation 7[8], that certain manufacturers (defind in Regulation 6) prepare a written report – commonly called "safety case" and to submit it to HSE. Schedule 6 of the Regulations specifies the information to be included in the safety case. In essence, the safety case is a demonstration that the manufacturers activity is being carried out safely.

The HSE guidelines on PESs, provide a design and assessment framework for the systematic examination of equipment in which a PES plays a role in the achievement of the required safety level. The guidelines address all aspects relevant to both random hardware and systematic failure causes. The method of approach adopted in the guidelines to both design and assessment (in particular the use of **Steps 1-6** and the **checklists** (Figures 7.3 and 7.4) could very usefully form the basis of the preparation of safety cases required by CIMAH. The adoption of a common framework, including the use of the checklists would be of benefit to both manufacturer and HSE. It should, for example, greatly facilitate communication between HSE and the manufacturer.

7.10 FUTURE DEVELOPMENT: GENERAL

Guidelines development will need to take place on both a generic and application-specific level and at both a national and international level. It is important to appreciate that in developing the guidelines, close contact was kept with our European neighbours. For example, a project that was run under the auspices of the European Commission [9,10], in which HSE was the project co-ordinator, allowed considerable cross fertilization to take place and influenced the development of the HSE guidelines. This project also influenced the framework adopted in the recently published Nordic guidelines [11].

The development of generic guidelines must be given a high priority because of the importance of ensuring that future application-specific guidelines development is based upon a rational framework with common underlying principles.

HSE would hope to act as a catalyst and provide guidance and help to industry to develop its own application-specific guidance based on the HSE guidelines. For example, HSE propose to follow the publication of the two documents with a number of publications which will further develop the guidance contained in "**PES 1**" and "**PES 2**" and be complementary to it. This further guidance will tackle key areas and thus ease the development, by industry, of application-specific guidance. Projects in hand include:

- Guidelines on the safeguarding of industrial robots – "Industrial Robot Safety" [12].
- An information handbook on hardware reliability data sources.
- Guidance on the design and assessment of emergency shutdown (ESD) systems for the process industries with particular reference to PES based systems.
- Development of hardware reliability criteria. Particularly that relating to the qualitative examination of safety related systems. This is to be developed as an adjunct to the **reliability** system element.
- Intelligent/"Smart" sensors. A study is being conducted to examine the safety advantages of these devices together with associated safety assessment criteria.

- Guidance relating to the **overall quality** system element. A high priority in this guidance will be given to the quality assurance aspects of PES design and manufacture.
- A publication giving an overview of the electrical interference aspects with particular reference to PES based safety related systems.
- Guidance on the testing and commissioning of PES based systems.
- Guidance on enhancing the safety integrity of hardwired systems using programmable controllers. This is primarily aimed at machinery safeguarding applications.

Guidance development of this nature, by HSE, will take place after discussion and consultation with industry and in some cases in close collaboration with industry. In the latter case this would apply particularly in the context of the application-specific guidance. It is important that guidance be available for both generalists and specialists in industry and this will also need to be addressed.

It is recognized that further guidance is required in the area of software used for safety related PESs. This is a very important area and a **Safety Related Software Study** is being undertaken for HSE, to cover the next five years, which should enable HSE to direct its resources in priority areas in the most effective manner. HSE propose to publish this **Study**.

7.11 FUTURE DEVELOPMENT: APPLICATION-SPECIFIC GUIDELINES

A major objective in producing *generically* based guidelines on PESs was to provide the basis for industry and others to produce their own application-specific guidelines. To this end HSE will develop *strategic guidelines* in specific areas (Section 7.10) in order to ease the task of industry in developing application-specific guidelines. Already a number of bodies are either actively producing such guidelines or actively considering what guidelines need to be developed. They are:

- The Institute of Gas Engineers.
- The Engineering Equipment and Materials Users Association (EEMUA).
- The Energy Industries Council (EIC).
- GAMBICA (The Association for the Instrumentation Control and Automation Industry in the UK).

7.12 FUTURE DEVELOPMENT: STANDARDS

It is important to ensure that guidelines development takes place taking account of the work going on within the standards organizations at both a national and international level and that guidelines that are developed nationally are progressed internationally. A major objective must be the achievement of international standardization.

Within the IEC, work is progressing on a number of topics which will have to be closely co-ordinated in order to minimize overlap and ensure that future guidelines in this area adopt the same underlying principles. In particular:

- Functional safety of PESs (**IEC/SC65A/WG10**);
- Safety related software (**IEC/SC65A/WG9**);
- Evaluation of system properties (**IEC/SC65A/WG8**);
- User guidelines for programmable controllers (**IEC/SC65A/WG6**);
- Electrical equipment of industrial machines (**IEC/TC44/WG3**);
- Electrical interference (**IEC/TC65/WG4**).

7.13 SUMMARY

- The HSE guidelines are intended to stimulate discussion on design and assessment aspects of safety related PESs.

- Elements of the safety strategy outlined in the HSE guidelines will be relevant to all PES applications, both existing and new but the extent to which they are applied in a given case will depend on the particular circumstances.
- The HSE guidelines attempt to set out the general principles applicable to the design and use of PESs in safety related applications. They apply to the *total configuration of safety related systems.*
- The framework in the HSE guidelines is relevant also to non-PESs.
- Future work on safety related PESs will need to proceed on both a generic and application-specific level.
- HSE will provide help and encouragement to industry and others in the development by them of application-specific guidelines.
- HSE will develop *strategic guidelines* in specific areas in order to ease the task by industry of developing application-specific guidelines.
- The work going on within the IEC on "Functional Safety of PESs: generic aspects" has important implications for the longer term developnent of standards covering safety related PESs — both generically and in the context of specific applications.
- There is a need to develop the "*total system environment*" model in order that all key features can be rationally considered.
- The concept of the safety related system (as used in the HSE PES guidelines) has important safety and economic benefits.
- The HSE guidelines provide a basis for the formulation of *safety cases* required under the CIMAH Reguations.

REFERENCES

1 Nimmo, I., "Lessons learned from the failure of a computer system controlling a nylon polymer plant", Institute of Measurement and Control, 4th Symposium on Microprocessor-based Protection Systems, London, December 1987.

2 Programmable electronic systems in safety related applications: "1 An introductory guide". (ISBN 011 8839136)*

3 Programmable electronic systems in safety related applications: "2 General technical guidelines". (ISBN 0118839063)*

4 Bell, R. and Robertson, S.S.J., "Programmable electronic systems in safety related applications: HSE guidelines", presented at an IEE colloqium on the "HSE guidelines" in London, June 9 1987.

5 Jones, P.G., "Safety considerations in the use of PES for the control of chemical plant", a paper based on "CHISA '87", Prague, Sept. 1987. Available from HSE Library, St. Hugh's House, Bootle, L20 3QA.

6 Brazendale, J. and Pearson, J., "Computer controlled chemical plant; Design and assessment framework", paper presented at Conference 11 titled "Preventing Major Chemical and Related Process Accidents", 10-12 May 1988; Queen Elizabeth II Centre, London.

7 Control of Industrial Major Accident Hazards Regulations 1984 (S.I. 1984/1902).*

8 "A guide to the Control of Industrial Major Accident Hazards Regulations 1984". Health and Safety Series booklet* HS(R) 21.

9 European project titled "Assessment, architecture and performance of industrial programmable electronic systems with particular reference to robotic safety", reported in proceedings of the "Programmable Electronic Systems Safety Symposium" held in Guernsey, UK 1986 and published as "Safety and Reliability of Programmable Electronic Systems", B.K. Daniels (Ed), Elsevier Applied Science Publishers, ISBN 1-85166-017-8.

10 Anderson, O., Bell, R., Meffert, K. and Vautrin, J.P., "European Collaborative Project on the Assessment of Programmable Electronic Systems", Journal of Occupational Accidents, Vol. 9, pp. 123-135, 1987.

11 "Personal safety in microprocessor control systems (A Summary) 1987", available from Nordisk-Ministerrad, Store Strandstraede 18, DK-1255, Copenhagen, Denmark.

12 "Industrial Robot Safety", HSE publication (ISBN 011 8839993), 1988.

***Note:** Copies of these documents available by post from HMSO, PO BOX 276,London SW8 5DT (Telephone orders: 01-622-3316).

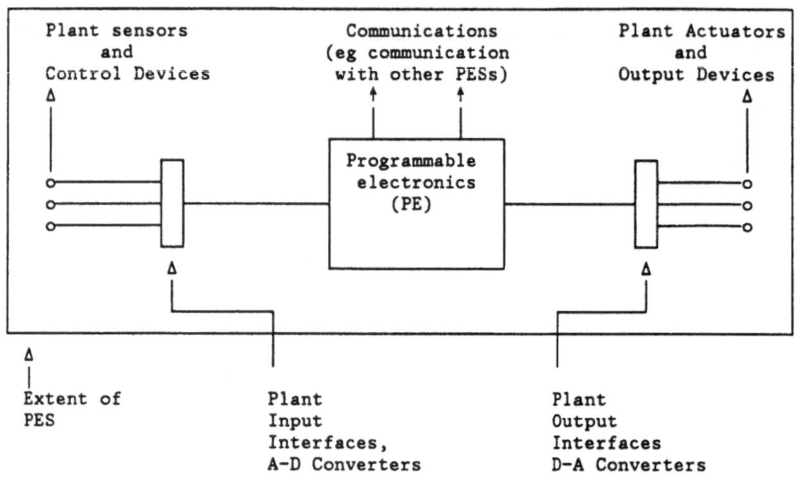

Figure 7.1. Basic PES structure.

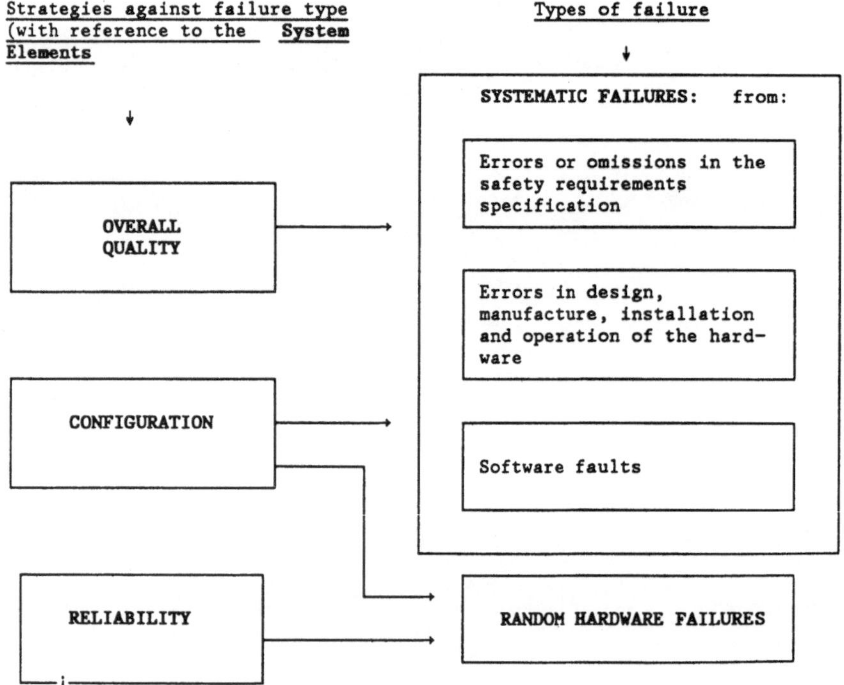

Figure 7.2. Design and assessment strategy.

(Used for detailed systematic approach in meeting *overall quality* system element)

Safety Requirments Specifications — Checklist No. 1

Hardware	Software (Checklists 10-13 apply to original software design)	Proprietary industrial programmable electronics (Checklists 14-15 apply to the use of proprietary industrial programmable electronics)
Hardware Specification — Checklist No. 2	Software Specification — Checklist No. 10	Embedded Software — Checklist No. 14
Hardware Design — Checklist No. 3	Software Design — Checklist No. 11	Application Programming — Checklist No. 15
Hardware Manufacture — Checklist No. 4	Software Coding — Checklist No. 12	
Hardware Test — Checklist No. 5	Software Test — Checklist No. 13	

Installation — Checklist No. 6

System Test — Checklist No. 7

Operations — Checklist No. 8

Hardware Maintenance & Modification — Checklist No. 9

Software Maintenance & Modification — Checklist No. 16

Figure 7.3. Organization and checklists.

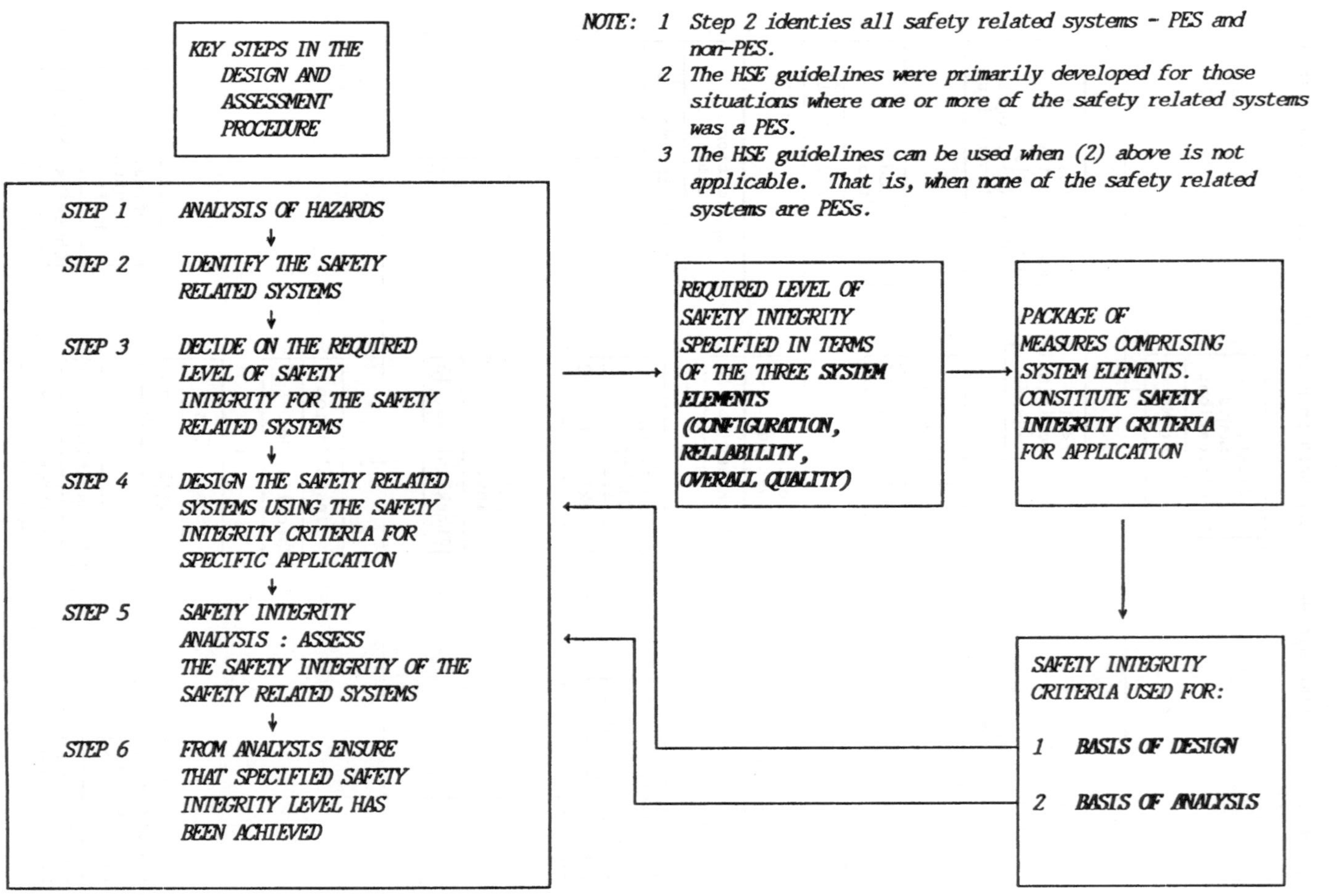

Figure 7.4. Overall framework: Design and assessment.

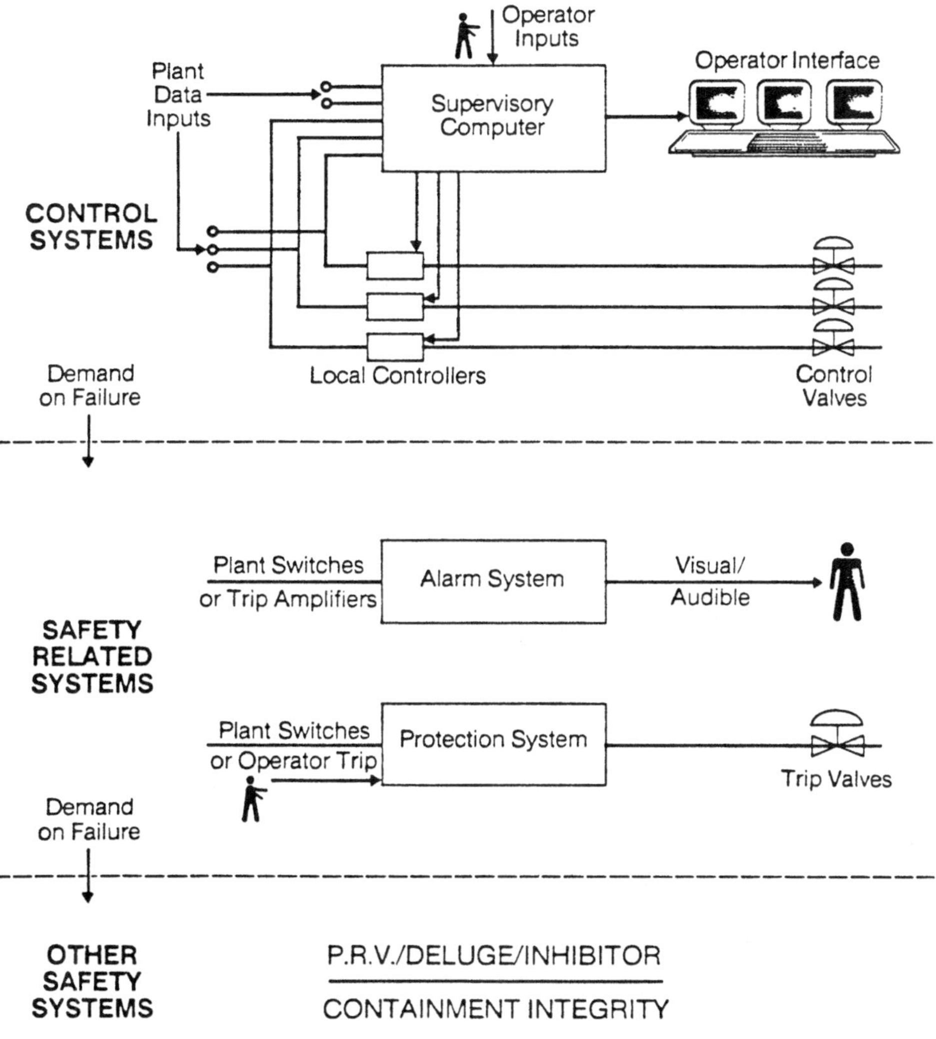

Figure 7.5. Typical configuration of control and protection systems for a chemical plant.

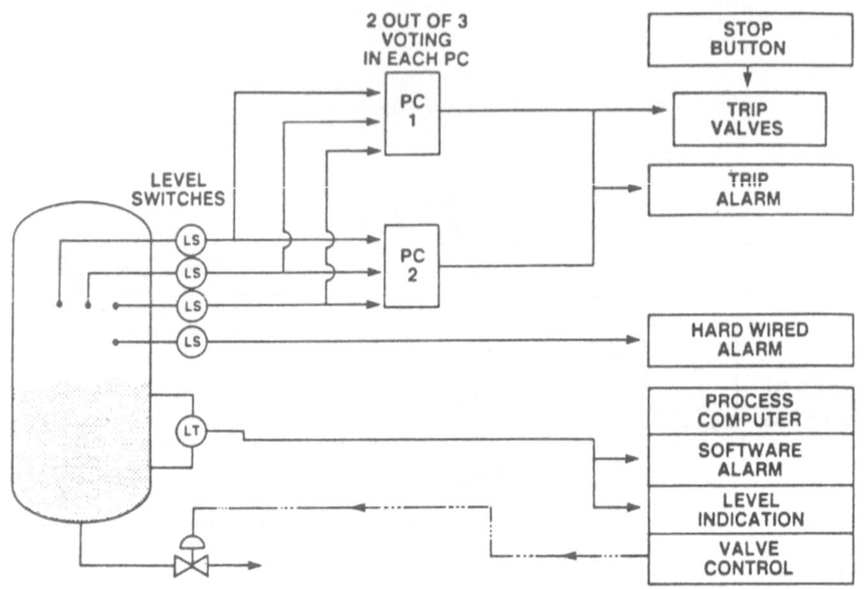

Figure 7.6.

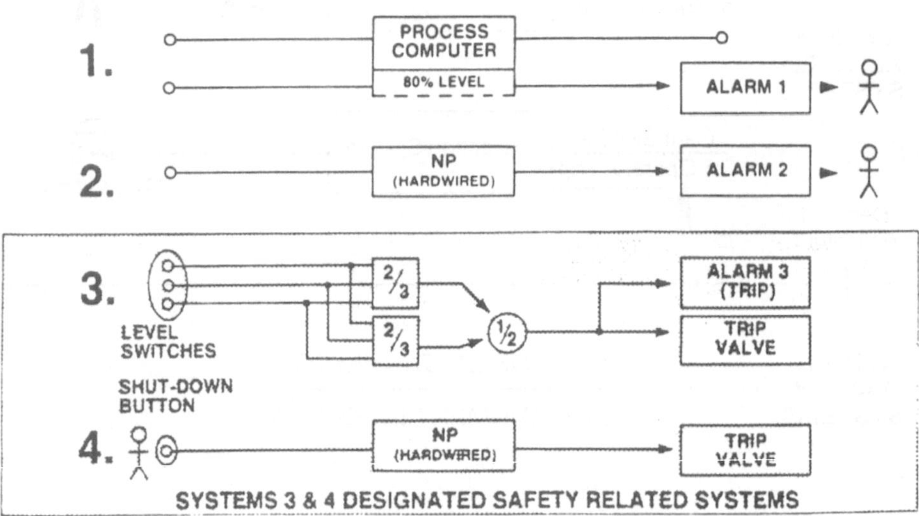

Figure 7.7. Configuration of control and protection systems.

Chapter 8

FAILURE-TO-SAFETY IN TURBINE-GENERATOR CONTROL

P.A.L. Ham
NEI Parsons Ltd., Newcastle Upon Tyne

8.1 INTRODUCTION

Modern Steam Turbine-Generators in the UK have a unit rating which is typically 500 to 660 MW. Even higher ratings are now in prospect, and units up to 1300 MW have already been installed by utilities abroad. This contrasts with the early 1950s, when 60 MW was the largest machine in common use. One consequence of this increase in unit rating is that the torque-to-inertia ratio of modern machines has increased considerably. The acceleration capabilities are therefore much higher and, as an additional factor consequent upon the higher ratings, all the components tend to be stressed more closely to their ultimate limits.

At the same time, economic considerations call for the maximum availability of the more efficient plant in order to minimize overall costs. Whilst this aspect receives much day-to-day attention, it is nevertheless an overriding requirement affecting both the design and operation that the plant shall operate safely. In practice this means that for all foreseeable single faults, the plant must react in a safe manner and in as many situations as possible it should be capable of continuing to generate power. In an ideal situation, the plant should be capable of continued safe operation in the presence of multiple faults, but as a final recourse it should always shut down safely.

8.2 MODES OF FAILURE

There are many possible failure conditions for the control systems associated with a Turbine-Generator. The approach to the design for each system must be conditioned by the degree of hazard associated with any failure, and also by the timescale in which the hazard may develop; some hazards are immediate whilst others may allow some minutes in which operator intervention is possible. The following categorization of faults is not exhaustive, but provides a good illustration of the scope and timescales which have to be considered.

(a) Overspeed condition

The centrifugal stresses in the rotating elements of a Turbine-Generator are inevitably a high proportion of the maximum allowable stress; at the same time, because of the relatively large inertias the stored energy is quite considerable. At the rated speed this is sufficiently high to result in very considerable damage if disintegration occurs and the fragments are not contained by the outer casing. Since the stored energy is proportional to the square of the speed, the situation can be exacerbated where disintegration occurs as a result of a loss of control following a disconnection from the grid; the speed rise before ultimate failure may be in excess of 70%, as appeared to be the case in the Uskmouth disaster [1].

This failure, which occurred in 1956, was the last of its kind in a British power station, and resulted in loss of life and injury as well as considerable damage. The source of the failure was traced to the presence of black iron oxide (magnetite) deposits in the control system actuator pistons, brought about by an unidentified leak of seawater into the lubricating oil. In modern terms this would be classed as a Common-Mode fault since it affected both channels of the overspeed protection system, and well illustrates the importance of this type of failure as has been discussed by Bourne et. al. [2]. In this incident the machine overspeed to destruction is approximately 13 seconds; due, however, to the much increased torque-to-inertia ratio of modern machines a figure of less than half this would now be appropriate. It will be appreciated that one of the most basic requirements in the control of Turbine-Generators is, therefore, to guard against overspeed.

(b) Failure of an auxiliary

A large Turbine-Generator is supported in its normal running by a number of Auxiliary systems, such as Seal Oil systems, Lubrication Oil systems, Cooling systems of various kinds and many others; there may be over twenty separate systems for a large machine. In most instances, failure is not immediately catastrophic in the manner of an overspeed, and there may be some minutes during which corrective action may be taken. However, the escape of steam can constitute a hazard, as can the release of lubricating oils which may ignite upon touching a hot surface. It is therefore necessary that protective systems shall operate automatically and not rely on operator intervention, if consequential damage to the machine is to be avoided.

(c) Elapsed life failures

In a highly-stressed machine it is inevitable that there will be a number of age-related factors which will place a limit on the life of the plant and could lead to ultimate failure which will be prejudicial to safety. These factors typically arise from three main sources :

(a) Direct Damage Mechanisms such as high temperature creep or fatigue cracking originating from material defects,

(b) Ageing Effects due to loss of strength with time due to metallurgical changes, and

(c) Mechanical Damage due to erosion, or distortions arising from creep and/or thermal stresses.

Many of these factors are accommodated by appropriate design methods, including factors of safety, but thermal stresses affect the whole Turbine to a greater or lesser extent and the severity is a function of the way in which the machine is operated. Whilst the steam chests and casings are subject to a certain measure of thermal stress, the most critical area is that of the HP and IP rotors; for this reason, most manufacturers have developed means for assessing rotor thermal stress, on-line, in order to allow the operator some criteria for running the machine to minimize the stress levels; many have extended the system to embrace analytical procedures for estimating the elapsed thermal fatigue life of the machine.

It will be evident from the considerations described above that it is essential, in the design and operation of large Turbine-Generators, to apply a wide range of safeguards in order to reduce the possibilities of failure to negligible proportions. This involves approaching the problem in a number of ways.

8.3 PROVISIONS TO ASSURE FAILURE-TO-SAFETY

The approach to the problem of assuring failure-to-safety is inevitably multi-levelled, with overlapping contingency protection and a measure of diversity. Whilst the application of the principles involved may well cover all aspects of the plant, there are certain areas which are particularly important.

(a) Control system design features

It is an essential feature of control systems for use in high-integrity applications that they are able to continue to operate safely in the presence of internal faults. This is generally achieved by means of Redundancy in which a number of parallel Channels of control are provided, and the application of this technique has been described in the context of Turbine-Generators [3]. The choice of control between the alternative signals from the Channels is achieved by a form of arbitration known as Majority-Voting, techniques for which have been described in a chapter of the book by Warwick and Rees [4]. There are a number of other properties, particularly relating to software, which may be important and systems embodying them have come to be known as "Fault Tolerant". A summary of the principles (particularly for digital systems) has been given in an IEE Review Paper [5]. The existence of a voting system allows any one channel which suffers a potentially dangerous failure to be rejected; systems of this kind are now standard for the Turbine and Generator controls of most manufacturers' machines, and a typical example using a Triplex Modular Redundant (TMR) system is shown in outline in Figure 8.1. It will be seen that full channel separation is maintained from the transducer inputs through the final output signals to the individual steam valve controllers. Each steam valve controller incorporates an individual majority-voter so that the system as a whole will continue in operation with any one channel out of service, and also one or more steam valve out of service.

It is generally desirable to structure the control system in a hierarchical manner such that the more basic (and safety-related) features are provided in a fully-redundant implementation, whilst higher-level functions which do not have the same safety connotations exist only in single-channel form. This approach has been adopted in a modern range of Turbine control systems [6] and is instrumental in achieving a suitable balance between failure-to-safety, complexity and overall cost.

The requirement for redundancy may be fairly clear in the case of complex microprocessor-based controllers, but the case is equally valid for more traditional systems. Thus in the control systems for Auxiliaries, such as the Lubricating Oil supply, it is normal to provide redundancy in the form of a back-up motor-driven pump which can be automatically brought into service should the running unit fail. In this particular instance, the consequences of failure are not likely to produce a safety hazard but carry such economic implications that a third standby pump running off a Station DC supply is often also provided as discussed in Section 8.3(c). Subsequently, this provides Diversity as well as Redundancy.

(b) Safeguarding of predictable fault modes

In any control system there may exist certain failure patterns which can be predicted by a detailed fault analysis. Thus, in a Turbine context it is evidently undesirable that the power source for the steam valve controllers should fail in a manner which leaves the admission valves unable to close and for this reason particular types of fail-safe system have traditionally been used. Operation of steam valves for large machines requires both a fast stroking capability as well as high force levels; this need is met by the use of powerful hydraulic mechanisms which, in the case of machines produced in the last 15 years, is typified by the electrohydraulic system illustrated in Figure 8.2.

Conventional hydraulic controllers used for industrial purposes frequently employ double-acting actuators; they commonly exhibit the characteristic that the output will remain stationary if the fluid supply pressure is lost. The use of a single-acting final stage to operate the steam valve confers failure-to-safety, since movement in the "valves closed" direction is provided not by fluid pressure, but by a return spring. This ensures that if the pressure falls, then all the steam valves will close, so giving a safe condition.

It should be noted that fail-to-dangerous conditions are often implicit in the failure of feedback transducers, since the loss of a feedback signal may well result in the control output being driven into positive saturation. To counter this situation, the valve control mechanisms utilize Linear Variable Differential Transformers (LVDTs) for position measurement which are continuously monitored for failure. In normal use the difference voltage on the two output windings represents the position signal; if no defects are present however, the sum signal assumes an approximately constant value and this can therefore be used with a level detector for monitoring purposes. The circuits which carry out the checking are located in the Local Driver Boxes also shown in Figure 8.2.

Electrohydraulic mechanisms rely on the use of an electrohydraulic servovalve to make the conversion between semiconductor and hydraulic systems. Although double-acting, the servovalve may be similarly set up to assume a safe condition in the event of failure. It is a requirement that the loss of all electrical control signals should result in closure of the steam valve; this is achieved by a mechanical "offset" set up within the servovalve such that the residual output flow will drive the mechanism in the "shut" direction. In some instances, such as in older machines updated to modern digital control by means of a special electrohydraulic converter, two low-pressure servovalves may be used in parallel and in these cases an unsymmetrical limitation of the drive currents ensures that there is always more control authority available in the "close" than in the "open" direction.

There are many instances in control system design where it is possible to provide safeguards against particular dangerous fault modes as in the examples quoted above; this applies not only to mechanical systems but also to analogue or digital techniques for dealing with software-based systems, and one which may particularly be noted is the use of a Watchdog Timer. In a system so equipped, every successful completion of a service routine resets a timer; if for any reason the timer fails to be reset, then after a short delay a safety action is initiated. This may variously be a shut-down or a reinitializing of the control processor.

(c) The provision of overriding safety mechanisms

Whilst the control systems associated with steam turbines are designed as far as possible to be fail-safe within themselves, reliance for ultimate safety is not necessarily upon them. Thus, in the particular case of overspeed a totally separate, redundant Overspeed Protection system is installed; this forms part of what is known as the Tripping System, a representative illustration of which is shown in Figure 8.3.

The tripping system is designed to shut down the Turbine-Generator as rapidly as possible if any one of a range of potentially dangerous conditions arise; this it does by releasing the fluid pressure in the steam valve control mechanisms, so cutting off the steam supply in approximately 200 milliseconds. Typical of the conditions for which an overriding shut-down action is considered essential are: Governor fault, low condenser vacuum, loss of lubrication oil, low Generator coolant flow and Generator excitation failure or protective relaying action.

It will be seen that the tripping system is redundantly structured and has both triplex and duplex sections which are interfaced by means of majority-voters. In the example illustrated, the main overspeed protection is by means of duplex centrifugal overspeed detectors known as "Bolts". These are entirely mechanical in action; their continued use in conjunction with a modern digital control system well illustrates the application of "Diversity", i.e. the use in a safety context of differing forms of technology. This approach minimizes the chances of faults of a similar kind arising at the same time in both the control and protection system.

8.4 FAULT DETECTION REQUIREMENTS

A system having fault-tolerant features, particularly where failure-to-safety is important, will have requirement for in-built fault detection. Where possible, this will be fully automatic but there will be some instances where interpretation is required and in such cases this will either be carried out by the operator or, increasingly in the future, by an "Expert System".

(a) Automatic on-line fault detection

It can be shown [2] that the security of a redundant control or protection system is significantly reduced if one channel has become unserviceable. It may be noted that the objective of redundancy is to cover continuity of operation by, in effect, masking faults; at the same time the statistical defect rate for a redundantly-structured system will of course be in direct proportion to (or perhaps somewhat greater than) the number of channels used; hence it is clear that control systems for use in a high-reliability or safety-related context must include monitoring systems which operate continuously to detect the development of internal defects.

A number of techniques are available for continuous on-line automatic fault-detection. The use of Modular Redundancy, in particular, allows the validation of signals at various stages in the control algorithm to be carried out by cross-comparisons between the channels. Validity-checking of all incoming and outgoing data should be carried out on the basis of ranges and possibly also rate-of-change of the variables. A particular advantage of digital systems may be the possibility of running on-line Proof Tests on a time-share basis with the main control program, so that apparent concurrency of both functions is achieved. Turbine-Generator control systems employ all these methods in various combinations.

(b) On-load testing

Certain types of defect may not be amenable to detection by automatic procedures of the type described above. These generally come in the category of "Unrevealed Faults" and can only be detected by causing the system to be exercised in some way. With Turbine-Generators, this requirement is met by a number of "On-Load Tests" in which safety-related systems are temporarily isolated and then required to operate in every way as if an emergency had arisen.

On-Load testing is applied at regular intervals to all the Tripping functions of the Turbine-Generator. This is particularly aided by the redundancy arrangements, where one channel can be tested whilst the others continue to operate as normally to maintain the safety of the machine. Full isolation can be obtained by means of mechanical/hydraulic interlocks to allow each of the overspeed bolts to be tested, either by simulating the centrifugal force by means of a hydraulic pressure, or else by actually increasing the machine speed under governor control. The operating point of the Bolts is always set well below that at which any hazard would occur.

Since it is possible that the build-up of contamination deposits in an actuator could prevent the rapid closure of steam-valve actuators in an emergency, the on-load tests include a controlled closure and re-opening of every steam valve. This practice, in common with several others, has arisen as a result of long and occasionally bitter experience spread throughout the power industry. Current practice also involves the archiving of performance traces, which allows any build-up of friction to be detected well ahead of any problem.

In some Turbines, particularly those associated with steam feed-pump drives, the size of the unit and notably the shaft diameters are not adequate to install a satisfactory Overspeed Trip Bolt arrangement of conventional type. For these (and a few larger machines) TMR Electronic Overspeed Trip systems are used as illustrated in Figure 8.4. These systems incorporate a separate processor which can carry out a sequence of tests so that the Trip point can be checked on each channel; the tests are carried out at intervals when initiated by the operator. In normal operation, however, a continuous monitoring operation is carried out in which the processor cross-compares the speed readings obtained from the three channels. In the event of failure of one channel, it is an important feature that the system reverts to an "any-one-from-two" mode; part of this is assured by the output relays being operated in an "energized-for-normal" manner.

8.5 CONDITION MONITORING

An increasing interest is evident in systems to assess the state of the Turbine-Generator and to not only detect faults but to form some kind of prognosis of developing fault situations. These come under the heading of Condition Monitoring, and include a number of aspects which inevitably are safety-related. It may be anticipated that in the future the development of CM systems [7] will make a significant contribution to the fail-safe performance of Turbine-Generators.

In this field, much attention has been focused on techniques for vibration monitoring, particularly in respect of Run-Down monitoring in which the various vibrational modes can be assessed as the machine speed slowly falls. This technique is most valuable in detecting the development of cracks in the Turbine or Generator rotors, which may be undetectable (particularly in the early stages) during normal running.

There are, however, many other techniques in addition to vibration monitoring. Typical of these is the "Core Monitor", which detects Generator insulation failure by identifying varnish particles released into the coolant gas, which has been available for some years [8]. More recent as well as search-coil detectors which can indicate the development of shorted-turns [9].

8.6 CONCLUSION

Turbine-Generators are a field in which the requirements of failure-to-safety have been prominent over many years, and the current situation has been described in outline. There is, however, continuous development of a whole spectrum of checks and monitoring procedures to identify actual and potential fault conditions of all kinds in Turbine-Generators and related plant, and this work will enhance the safety-related performance of future machines.

ACKNOWLEDGEMENT

The Author thanks the Directors of NEI Parsons Ltd., for permission to publish this Paper.

REFERENCES

1 Lindley, A.L.G. and Brown, F.H.S., "Failure of a 60-MW steam turbo-generator at Uskmouth power station", Proc. I.Mech.E., Vol. 172, No. 17, pp. 627-642, 1958.

2 Bourne, A.J., et. al., "Defence against Common-Mode Failures in Redundancy Systems", SRD/R/196, UKAEA, Culcheth, Warrington, 1981.

3 Ham, P.A.L., "The Application of Redundancy in Controllers for High Capital Cost or High Integrity Plant", IEE Conference: "Trends in On-Line Computer Control Systems", Pub. No. 208, pp. 181-191, 1982.

4 Warwick, K. and Rees, D. (Eds), " Industrial Digital Control Systems", IEE Control Engineering Series No. 29, 2nd. Edn., Chapter 12, pp. 242-266, Peter Peregrinus Ltd., London, 1988.

5 Depledge, P.G., "Fault Tolerant Computer Systems", IEE Proc., Part A, Vol. 128, No. 4, pp. 257-272, 1981.

6 Ham, P.A.L. and Green, N.J., "Developments and Experience in Digital Turbine Control", IEEE/PES Trans. Paper No. 88 WM 236-2, 1988.

7 Ham, P.A.L., "Trends and future scope in the monitoring of large steam turbine generators", IEE Proc., Vol. 133, Pt. B, No. 3, pp. 164-167, 1986.

8 Wood, J.W., "Condition monitoring of turbo-generators", Electronics & Power, Vol. 28, 10, pp. 682-685, 1982.

9 Wood, J.W. and Hindmarch, R.T., "Rotor winding short detection", IEE Proc., Part B, Vol. 133, pp. 181-189, 1986.

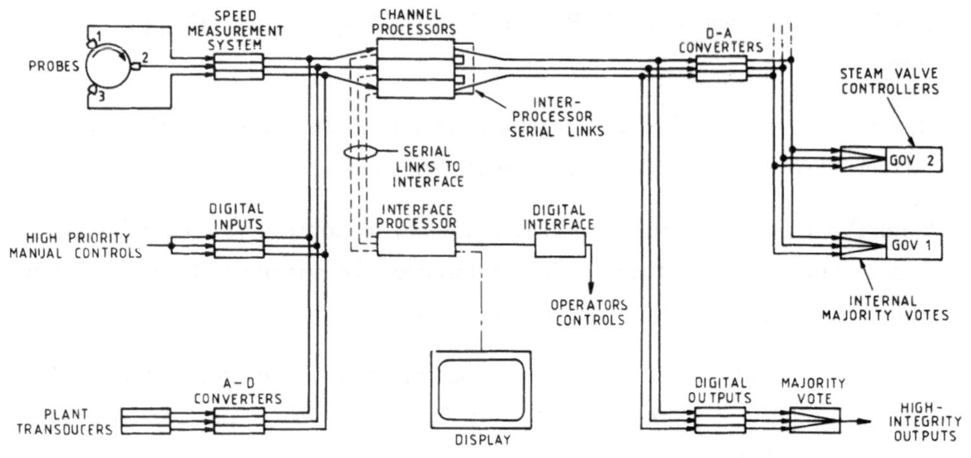

Figure 8.1. Examples of a "Triplex Modular Redundant" (TMR)

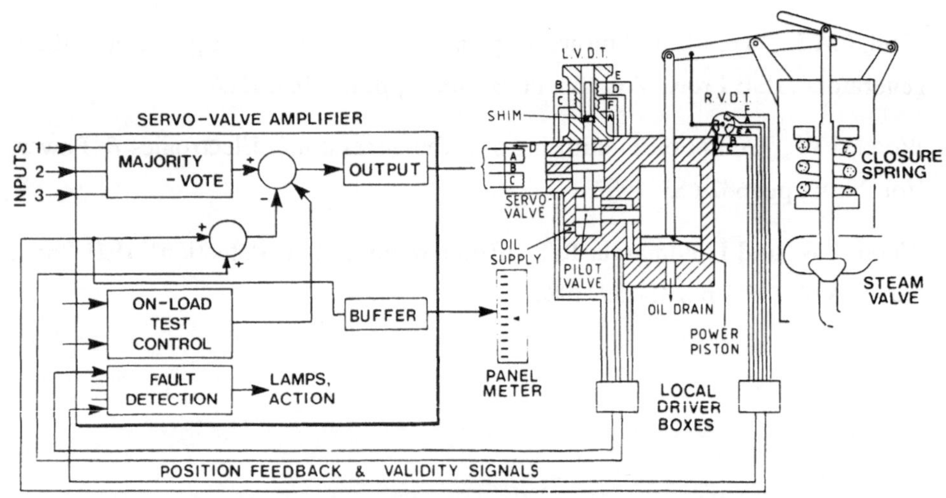

Figure 8.2. Fail-safe Turbine steam valve control system.

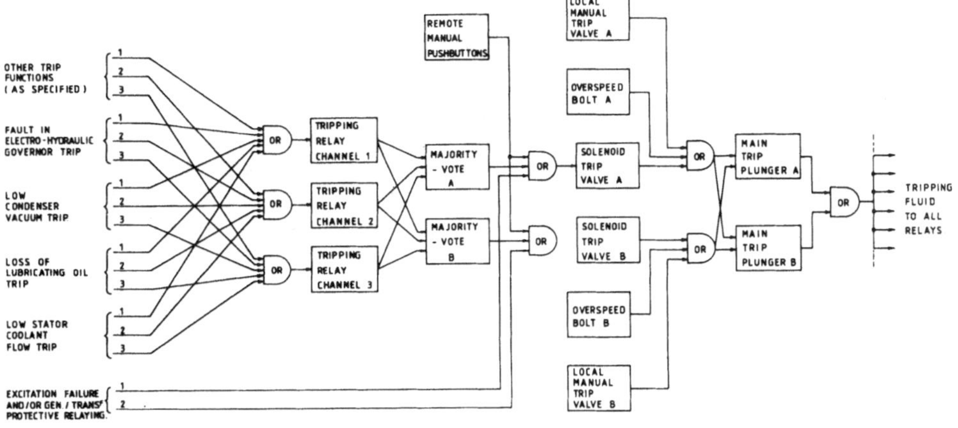

Figure 8.3. Logic diagram for a turbine tripping system.

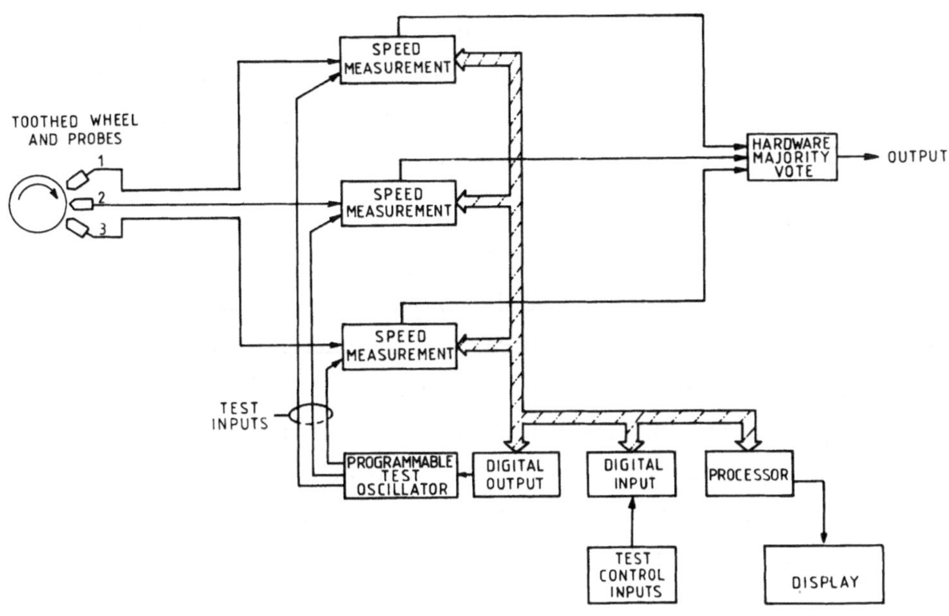

Figure 8.4. Electronic overspeed trip system showing test facilities.

Chapter 9

EXPERT SYSTEMS FOR MONITORING PROCESS CONTROL

Dr. Robert W. Milne
Intelligent Applications Ltd

9.1 INTRODUCTION

Process control and monitoring is becoming increasingly automated. In today's environment of PID and PLC controllers, large process control systems and distributed control systems, the computer is involved in many phases of process control. Complementary to that trend is the introduction of Artificial Intelligence [21] for interacting with the process control system in several strategic ways [7, 10]. In this chapter, we discuss one of these opportunities, process monitoring, in more detail and discuss its relationship with existing control systems.

There has been some exploratory work on the role of Artificial Intelligence in process control itself [2, 20]. Several tool kits have been developed to assist with this process [8, 18]. In this approach, Artificial Intelligence is usually used to replicate or replace the control of the process itself. As an emerging technology, it was necessary to experiment with Artificial Intelligence to understand how it could help and support process control. However, these experiments have largely led to trying to rediscover many of the main ideas of process control. As successful as they are and in the new ideas that were presented, by and large the techniques were not as effective as existing PID controllers.

Based on these experiments and a broader view of Artificial Intelligence in the industrial market place, we believe that Artificial Intelligence should only be used in monitoring process control systems today and in assisting with the operator to understand why the process controller has failed and how to get it back working as quickly as possible [12]. Many years of control system development have resulted in a state of the art of process control. Although Artificial Intelligence provides many powerful means for data abstraction and easy modification of instructions, they still do not address the depth of control needed [1]. Implementations of current techniques tend to be far too slow for many real time applications. Current hardware is too expensive to replace a low cost process control system and yet provide similar performance. The need to justify higher cost hardware has caused our concentration on very large scale expensive projects. This brings great danger since a relatively new technology should be tried in small ways at first.

9.2 BENEFITS OF THE EXPERT SYSTEM

Before a new technology is introduced, it must provide clear benefits to address needs in the existing process control. Process monitoring systems directly address many limitations inherent in the requirement for a human to monitor the system. Although the process parameters are relatively simple to see at a glance, it is rare that a human operator can stand continuously over the system, monitoring all the parameters. Very small systems normally do not justify constant human attention. Very large systems require constant human attention, but often contain thousands of parameters rather than hundreds.

Software has been developed to provide monitoring before. This software often requires an experienced person to develop and is very inflexible. A process control engineer does not want to spend time learning how to develop complex software, rather he wants to be able to adjust set points quickly and easily, get a big picture of the operation of the system, or provide changes in the interpretation of the process. Many of the monitoring situations require the knowledge of the process that the engineer has to be used in conjunction with the process parameters. An expert system for process monitoring addresses these points directly. Rather than a person being required to monitor, the expert system has the capabilities to embed his intelligence into evaluating situations and recommending courses of action.

In order to develop a monitoring system, the knowledge used by an experienced engineer is often necessary. The problem is how to represent this knowledge easily in a computer and how to manipulate it. This is precisely what the expert system rule language provides. There is also a very strong need to be able to adjust the software rapidly and easily. Conventional computer programming languages make this far too difficult to be practical on a wide basis. Expert systems, with their interactive editors and English like rule languages, provide a very simple means to meet this objective. Although we do not pretend the programming changes are trivial at this time, they are certainly as easy as possible to rapidly build up rules of interpretation or modify the way the system is working. The expert system provides a high level abstraction of the process which allows the engineer a much easier means of controlling the interpretation and monitoring program.

9.3 PROCESS MONITORING

Let us assume a current process control installation, in place will be PID and PLC controllers as appropriate, complete data acquisition systems and either a central or distributed computer system containing a dynamic data base with knowledge of the state of the process. The dynamic data base is used to constantly reflect the current state of all relevant process variables. The process controllers themselves are responsible to maintain specific parameters within specific alarm or set points. For example, the water pumping system will be responsible to keep the level in the tank at a specific depth. Dedicated hardware is used to ensure that this happens in a timely and accurate fashion.

The appropriate role for Artificial Intelligence is to monitor the information contained in the dynamic data base and the results of the controllers to provide either an early prediction of failure or provide a rapid means by which the system can be corrected if a problem does occur. Rather than use Artificial Intelligence to ensure the tank level stays at a particular depth, Artificial Intelligence should be used to understand that it is about to fail at this task and rapidly to understand the total process and what must be done to correct the problem. This involves three primary areas: prediction, understanding the state of the process and interpreting it and finally, meta or high level process control.

If we assume a simple example where a pump is being used to keep water level at a constant depth, we can also monitor the pump power draw or, in other words, the effort required by the pump to keep water at that level. While the PID controller turns the pump on and off as needed, the expert system monitor can monitor both the water level and the effort required by the pump. If a trend of the effort by the pump shows that the effort is increasing fairly rapidly, then a prediction can be made that there is a problem and that it may not be possible to continue keeping the water at that level [14]. Before the process controller fails to be able to maintain the water level by perhaps a pump failure, it should be possible to take other corrective action. If the pump has failed, it should be possible for the diagnosis system to monitor a wide range of state variables for the process and immediately instruct the operator to inspect the pump or inspect the inlets which may be causing a reduced flow rate.

Another major consideration in process control is the safety and security of the control system. For many commercial applications, it is necessary to guarantee that the process control system will work reliably and consistently. Introducing Artificial Intelligence technology into the control group provides serious problems in meeting this goal. By using it as a monitoring and backup system, many of these difficulties are avoided.

In the following sections we look at several major issues in the use of Artificial Intelligence for monitoring and how these can be addressed.

9.4 SPEED CONSIDERATIONS

One of the greatest demands that a process places on an Artificial Intelligence system is being able to respond in a guaranteed time, that is in a real time fashion [13]. The technology for Artificial Intelligence is still new enough and the speeds of the rule bases uncertain enough, that this problem should be avoided. In other words, Artificial Intelligence should be used in situations where the speed is not a constraint. This may imply that the system needs to go very fast. It is one problem to build a monitoring system for process control, it is quite a separate problem to build a time management element so that it knows how much time is available, how long the calculation might take place and adjust its actions accordingly [5, 9]. This area is new enough and difficult enough that this time management problem should be avoided.

Time itself however is a very important consideration. The timing of events and the pace at which they happen should be used extensively by the monitoring system to make predictions and assessments of the current situation. All incoming data should be time stamped for use in both the trending mechanism and to allow history systems to examine what has been taking place and the sequence of events. This sequence is particularly important for alarm processing.

9.5 THE USE OF HISTORY

Histories can be used in two major roles within the monitoring system. These can be thought of as local and long term. If a process fails immediately the local history should be used to help assist in a diagnosis. In development of rules and monitoring systems in the long term, long histories should be kept to disk and be available for post processing.

When a process control system has failed to maintain the set points desired, normally the current set of measurements are meaningless in that they reflect the system in a failed state. It is very important to be able to look back recently in time and find out which parameters deteriorated [19]. It can be very important to know that the current draw of the motor went up very rapidly immediately before the water level dropped. It could also be just as important to know that the water level dropped with no change in demand on the current. These could reflect a difference in fault between the inlet becoming plugged or a sudden hole or opening of the outlet for the tank.

In the systems we have developed, a rolling history is kept of a small number of most recent events, so that at any time it is possible to look back in history several minutes to find out how the system has been changing. The use of this data is similar to the ideas behind the PID algorithm, but it has been extended to support the diagnosis task specifically. It may be the case that many process engineers are not sure what to do with this recent time history since for the first time they have a snapshot of what happened to the system as it deteriorated, rather than after it has failed. With many systems, the exact methods of failure and what might happen immediately before a problem are unknown.

Keeping track of the significant process variables in a long term data file provides a very effective means for helping to develop the monitoring and diagnosis rules and for testing the rule base. Rather than hook an unproven expert system up to the process control system, one should first log a large amount of process parameters. In an off-line manner, the rule base can then be developed and tested on the actual data as it would have occurred on the plant. This provides a very important way to prove the capability, at least with historical runs. In our system, the data can be stored to disk or sent live to the rule base. As a result, it will behave exactly the same regardless of the input methods.

One of the important benefits of a rule based monitoring system is it can also be viewed as an intelligent data logger. Many data logging applications require all the parameters to be recorded or none at all. Through the use of the expert system shell, it is possible to monitor one or two variables, say the tank input and output flow rate. As long as they are within normal parameters, it is not necessary to record any data. If the flow rate on the input starts to go too low, we can have the pumping parameters and engine motor parameters stored to the disk in case a problem occurs. If it is detected that the process is struggling to maintain limits, then all the process parameters can be stored at a very fast sampling rate to the disk so that in the case of failure it is possible to go back and examine the history. This data logging as a result can be very flexible and controlled to occur only in the situations which will provide benefit.

9.6 TRENDING AND FORECASTING

So far we have discussed monitoring the process itself and providing the capability to diagnose problems immediately after they occur. The real pay-off for the monitoring system arises when the opportunity exists to predict a problem before it actually occurs. In Annie, it is possible at any point to take a trend on a particular data channel for the most recent values. The trend is based on the assumption that the system remains steady for a long time and then deteriorates fairly rapidly. An appropriate curve is fit depending on how the system is expected to change and the expert system is able to take advantage of the parameters of the curve fit, the quality of the curve fit and its first and second derivatives. As a result, in addition to just knowing the current value of a parameter, we can also find out whether it is rising or falling and how fast that is changing. The standard deviation is vital in evaluating whether this curve fit is meaningful and can be used to temper the probabilities of the forecasts. There are two fundamental forecasts that are required.

Given we have a particular alarm limit for a parameter, the key question is to ask, "how long until that parameter will exceed the alarm?" If we have a large process, it is important to know how many more hours the process can run until the tank must be cleaned. As a result, we can schedule a shut down and maintenance activity, either during the next overnight period, over the next weekend, until the next annual maintenance or perhaps in the next 45 minutes. Looking at smaller timescales, if it is known that a particular parameter is rapidly deteriorating and there are only several minutes to take corrective action, the expert system would be needed to make advice to the operator immediately, so he has time to correct the problem. Rules can be developed to constantly watch for parameters and give the operator say 5 minutes' warning of any known problem.

The other forecast is to find out what the values may be at any particular time. If we plan to shut the process down and clean the tank on Saturday, we may want to discover what the acceptable limits will be so we can determine whether other elements of the system need attention at the same time. If we know a process will only be running for the next 6 hours but the prediction is that it will exceed the threshold in 5 hours, we may want to know exactly what the value will be in 6 hours, so we are able to make an intelligent decision as to whether to shut it down or let it keep running through the alarm limit. It should be noted in many cases the process controller itself will allow no change in the parameter and as a result the forecast will not provide meaningful data. The forecasting is intended for the situations where parameters will deteriorate and predictions are possible.

9.7 SUPPORTING DATA ACQUISITION

The expert monitoring system must receive data either direct from the data acquisition [1] system or from the dynamic dictionary at regular intervals or as required. Fundamental to every application is the desire not to build a second data acquisition system merely to support the expert system. This means that process monitoring systems must be tightly integrated with existing process control systems. In some instances it is not possible to interfere in any way with the process control system and the monitoring system must tap into the termination panels and use its own separate A/D conversion equipment.

In other cases it is possible for the process controller to set up an independent process to receive requests from the monitoring system and provide replies. Where the process control computer has spare process time to do this, this provides a very satisfactory solution. Our experience, however, has been that process control systems were not designed to respond to external systems. This means that often a major expert systems effort can be prevented because of the constraints of the existing data acquisition system and the caveat that no additional data acquisition should be added. This factor will limit many current applications but wants to be seriously considered as new process control systems are being developed.

Finally, for many larger process control systems, such as those on Digital Equipment VAX computers, it is possible to have the expert system running directly accessing the dynamic data base. In this way, there is no interference with the real time data acquisition and no special means are needed to provide the data to the expert system.

At the other extreme, many expert systems are being developed requiring human or manual input. This is clearly not practical in these application areas, so data acquisition becomes a very important issue. The factors dominating data acquisition are largely not Artificial Intelligence specific problems but reflect the kind of engineering concerns that an Artificial Intelligence effort must encompass. The major concern we have is the great need for interaction between process control providers and the expert system development committees. In many situations, it is not possible to develop an expert system application without intense co-operation from the process control system developer.

9.8 META PROCESS CONTROL

In this chapter we have discussed several aspects of using expert systems to monitor process control systems. Rather than try and replace the functionality of the current control system, they monitor the process as it occurs, providing a constant understanding of the process and when a problem is about to occur or has occurred, rapidly can provide means by which the problem can be rectified [6]. This extra level of understanding is vital in the monitoring system. Rather than being focused on an individual parameter and how to keep it at a constant level, this system is trying to understand the process as a whole and how it might have failed.

This work points to the next step in using Artificial Intelligence within the process control domain. The understanding of the process at a higher level than monitoring individual parameters provides a method for meta level control [3, 4, 17]. A good process engineer can examine the big picture of a process and assess what other actions should take place to optimize the overall process. Control of the process of process control will be the next step in using Artificial Intelligence. This task normally requires considerable expertise, understanding and judgement on the part of the experienced process engineer. The knowledge that is used in this task can now be captured through expert systems and focused in this way. Future developments of our own systems, and I believe systems throughout the industry, will focus more on this task.

REFERENCES

1 Barr, A. and Feigenbaum, E., "The Handbook of Artificial Intelligence", William Kaufmann Press, Ca. 1981.

2 Clocksin, W. and Morgan, A., "Qualitative Control", Proceedings of ECAI-86, Brighton, England, July 21-25 1986.

3 Dague, P., Raiman, O. and Deves, P., "Troubleshooting: when modeling is the trouble", AAAI'87 Proceedings, Vol. 2, pp. 600-605, Seattle, Washington, 13-17 July 1987.

4 De Kleer, J., "Causal and Teleological Reasoning in Circuit Recognition", MIT AI Lab. Technical report, September 1979.

5 Forbus, K., "Qualitative Process Theory", MIT Technical Report No. 789, July, 1984.

6 Genesereth, M., "The Use of Design Descriptions in Automated Diagnosis", Stanford Heuristic Programming Project Memo HPP-81-20, January 1984.

7 Harmon, P., "Expert Systems Strategies", Cutter Information Corp., Vol. 2., No. 8, August 1986.

8 Kaemmerer, W. and Allard, J., "An Automated Reasoning Technique for Providing Moment-by-Moment Advice Concerning the Operation of a Process", AAAI'87 Proceedings, Vol. 2, pp. 809-813, Seattle, Washington, 13-17 July 1987.

9 Mavrovouniotis, M. and Stephanopoulos, G., "Reasoning with Orders of Magnitude and Approximate Relations", AAAI'87 Proceedings, Vol. 2, pp. 626-630, Seattle, Washington, 13-17 July 1987.

10 Milne, R., "Diagnostics and Machine Health – On-Line Expert Systems", Expert Systems in Process Control Seminar, London, 14th October 1987.

11 Milne, R., "Artificial Intelligence for Online Diagnosis", IEE Proceedings, Vol. 134, Pt. D, No. 4, July 1987.

12 Milne, R., "Strategies for Diagnosis", IEEE Transactions on Systems, Man, and Cybernetics, Vol. SMC-17, No. 3, May/June 1987.

13 Milne, R., "On-Line Artificial Intelligence", The 7th International Workshop on Expert Systems and Their Applications, Avignon, France, 13-15 May 1987.

14 Milne, R., "Artificial Intelligence Applied to Condition Health Monitoring", Chartered Mechanical Engineer, Mechanical Engineering Publications Ltd., Vol. 33, No. 5, May 1986.

15 Milne, R., "Fault Diagnosis and Expert Systems", R. W. Milne and B. Chandrasekaran, The 6th International Workshop on Expert Systems and Their Applications, Avignon, France, 28-30 April 1986.

16 Milne, R., "Fault Diagnosis Through Responsibility", Proceedings of the Ninth International Joint Conference on Artificial Intelligence, pp. 424-427, Los Angelos, August 1985.

17 Milne, R., "The Theory of Responsibilities", SIGART NEWSLETTER, No. 93, pp. 25-30, July 1985.

18 Moore, R., "Expert Systems in Process Control: Applications Experience", 1st International Conference on Applications of Artificial Intelligence to Engineering Problems, Southampton, England, 15-18 April 1986.

19 Reiter, R. and de Kleer, J., "Foundations of Assumption-Based Truth Maintenance Systems: Preliminary Report", AAAI'87 Proceedings, Vol. 1, pp. 183-188, Seattle, Washington, 13-17 July 1987.

20 Scarl, E. A., Jamieson, J.R. and Delaune, C.I., "Process Monitoring and Fault Location at the Kennedy Space Center", Sigart Newsletter, No. 93, July 1985.

21 Winston, P., "Artificial Intelligence", Addison-Wesley Press, 1984.

Chapter 10

ROBUST FAULT DIAGNOSIS IN DYNAMIC SYSTEMS

R. J. Patton and S. M. Kangethe
University of York.

10.1 INTRODUCTION

Many schemes for fault detection in dynamic systems have been introduced in recent years which use analytical or functional redundancy [1,2,3,4]. Fault tolerance especially those in which faults could result in a catastrophic event is usually achieved through hardware redundancy. To enable normal operation to continue in the presence of a fault, critical hardware, such as sensors and computers, are repeated in triplex or quadruplex redundancy schemes. Repeated elements are usually distributed spatially around the system to provide protection against localized damage. Such schemes operate by comparing outputs from repeated hardware in majority voting so that software/hardware faults may be detected and isolated. The major problems encountered with hardware redundancy, and those problems which have led to investigations into other forms of fault detection, are the extra weight and the space required to accommodate this, which could be used for example for more mission-orientated equipment.

To overcome these problems, and improve the overall system reliability and performance with the high integrity control systems (now required in many aerospace applications), recent work has concentrated on the use of analytically redundant relationships between dissimilar outputs of the same system. The majority of such methods are based on the use of state estimation. The process on which the fault/no fault decision takes place is based on the comparison of redundant data. The difference between redundant data (usually termed the residual) in an ideal situation is a measure of the performance of the sensor outputs under consideration. In all cases the decision process applied to residuals is a threshold criterion in which a fault is signalled when a threshold is exceeded. The factor common to all state comparison techniques is the use of some form of state estimator to produce estimates of part or all of the system state [1]. The state estimates may be compared, either with available system states or against estimates generated from different measurement subsets.

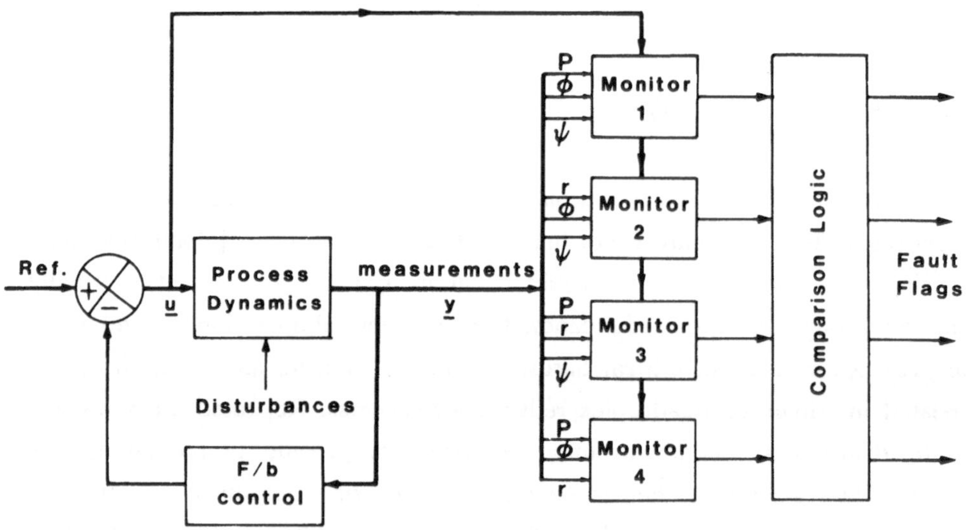

Figure 10.1. I.F.D. Monitor scheme.

The main problem associated with many Instrument Fault Detection (IFD) schemes lies in the certainty required in modelling the physical system, in design and implementation stages. Most techniques make certain assumptions about the physical system and/or the modes which may not be accurate, and may degrade the performance of the monitor scheme from that which was predicted theoretically. These limitations can cause linear model-based designs to be inadequate for many real applications. Sensitivity to input-induced parameter variations cause uncertain errors between redundant state estimate vectors and, in an IFD scheme these errors would cause false signalling of a fault. The solution proposed by previous investigators is to widen the threshold band for the fault-detect signal and so reduce false alarms. This also increases the time taken to detect a fault. The increased detection time may prove unacceptable in many fault-critical systems, such as those found in aerospace applications.

Any information available concerning uncertain dynamics should be used at the design stage to improve the performance of the IFD monitor. For many applications frequency domain or sensitivity information is available, or may be inferred from the desired control characteristics, about parameter variations and disturbances. From a consideration of the non-linear process a scheme has been developed which uses the above information in the design stage of an observer based on the output zeroing of scalar combinations of elements of the observer **e** error vector. This together with a scheme of decentralized observers produces an effective solution to the problem of fault detection and isolation for dynamic processes in which modelling uncertainties are significant [for further details the reader is referred to reference 3, Chapter 4]. Figure 10.1 shows a fault diagnosis scheme comprising a bank of observers with dissimilar measurements used as input. By taking all permutations of the available measurements p at a time the fault in a sensor can be detected quickly and isolated through the available redundancy relationships.

10.2 PROBLEM SPECIFICATION

The assumption made prior to IFD design is that the system can be represented by a linear model, a poor assumption for most real applications. It is instructive to consider a representation of a real system which is subject to parameter variations, and disturbances applied to a linear observer. Let the complete dynamic system be represented by:

$$\dot{\underline{x}} = \begin{bmatrix} \underline{x}_0 \\ \hline \underline{x}_u \end{bmatrix} = \begin{bmatrix} A_0 + \Delta A & A_c \\ \hline A_{u1} & A_{u2} \end{bmatrix} \begin{bmatrix} \underline{x}_0 \\ \hline \underline{x}_u \end{bmatrix} + \begin{bmatrix} B_0 + \Delta B \\ \hline B_u \end{bmatrix} \underline{u} + \begin{bmatrix} D_0 \\ \hline D_u \end{bmatrix} \underline{v} \tag{2.1}$$

$$\underline{y} = [C_0 \; C_u] \, [\underline{x}_0 \; \underline{x}_u]^T$$

$$y = [\,C_0 \quad C_u\,][\,x_0 \; x_u\,]^T$$

The matrices A_0, B_0 form the linearized plant to be modelled in the observer. ΔA, A_C, A_{u1}, A_{u2}, ΔB, and B_u represent the parameter variations and plant uncertainty which describe the time-varying dynamics of the plant, the subscript u denoting unmodelled effects. The subscript c denotes coupling to unmodelled dynamics. D_0 and D_u are the system disturbance distribution matrices, whilst $\mathbf{v}$ is the disturbance vector. C_0 is the pxn output matrix of the linearized plant.

The nominal linear system (A_0, B_0, C_0) is to be used in a linear observer with feedback gain K. The full-order form of the state observer used as the linear (nominal) model is assumed to replicate the actual process. For a more detailed treatment of the theory and design of the state observer the reader is referred to the original paper by Luenberger [5] or to a text book on modern state variable control theory (e.g. Blackman [6]). It will suffice here to provide the general form of the observer state estimation equation. The observer estimate $\tilde{x}_0$ of the partial state vector $x_0 \in R^n$ is given by the solution to the following:

$$\dot{\tilde{\underline{x}}}_0 = A_0\tilde{\underline{x}}_0 + K(\underline{y} - C_0\tilde{\underline{x}}_0) + B_0\underline{u} \tag{2.2}$$

If this observer is applied to the nominal system (denoted by the 0 subscripts) then the state estimate error is dependent only upon the initial error between the actual state and the observer estimate. When the linear observer is applied to the actual plant the impulse response of the estimator error is convolved with disturbance terms arising from uncertain effects. When driven by the actual plant output, the estimation error dynamics are as follows:

$$\dot{\underline{e}} = (A_0 - KC_0)\underline{e} + \underbrace{[\Delta A \mid \Delta B \mid A_C - KC_u \mid D_0]}_{E}\;\underbrace{[\underline{x}_0 \mid \underline{u} \mid \underline{x}_u \mid \underline{v}]^T}_{\underline{d}^T} \tag{2.3}$$

where $\mathbf{e} = \tilde{x}_0 - x_0 \in R^n$, and K is the nxp observer gain matrix. A_0 is an nxn matrix and C_0 is a pxn matrix. The estimation error equation now has an extra input vector which will have the effect of driving the error away from its zero steady-state during disturbances and plant parameter variations. The observer error dynamics may now be completely described by:

$$\dot{\underline{e}} = (A_0 - KC_0)\underline{e} + E\underline{d} = D_0\underline{e} + E\underline{d} \tag{2.4}$$

where $\mathbf{d}$ is an *unknown* disturbance input vector and E is an *unknown* distribution matrix. It is important to use (at design) frequency domain or sensitivity information about disturbances in an attempt to make the estimation error insensitive to effects of the unmodelled dynamics. This approach can be applied to the design of a robust (disturbance discriminating) observer in an attempt to decouple the effects of the disturbance term E $\mathbf{d}$.

10.3 ROBUST FAULT DETECTION

(a) Introduction

An observer is robust in this context if the time response of equation (2.4) remains de-coupled from uncertain disturbance terms E $\mathbf{d}$. A robust design can be achieved by using the theory of invariant subspaces provided originally by Rozonoér [7] and discussed further by Wang [8], Cruz and Perkins [9], Wonham [10], Willems [11] and Mita [12].

There are *two* main approaches to minimum disturbance sensitivity:

(a) The design of a $[H, D_0]$-invariant subspace (by suitable design of matrix H)

(b) The design of a $[D_0, E]$-invariant subspace

An approach based on (a) uses the concept [7, 12] that a system measure has *zero sensitivity* to a particular mode and modal parameter structure if it is also *unobservable* to this mode. A similar definition for controllability relates zero controllability with zero sensitivity and leads to approach (b) above. A combination of (a) and (b) can also be used to generate a system measure which can be completely uncontrollable by disturbances. The current work is based on the [H, D_0] – invariant subspace approach (a), however the designs can be shown to possess insensitivity to the uncertain disturbances given by E **d**.

(b) Continuous-time analysis

In order to examine the invariant conditions it is necessary to consider the *complete* observer dynamics (2.4) in continuous-time form as:

$$\dot{\underline{e}} = A_0\underline{e} - K\underline{e}_y + E\underline{d} \tag{3.1}$$

$e_y = y - \tilde{y}$ is the *measured* estimation error and can be looked upon as a control signal which drives the estimation error state towards an *invariant subspace* manifold *asymptotically*.

The invariant subspace motion occurs when the estimation error state lies simultaneously in each of the hyperplanes S_i, for $i = 1, \ldots, p$ given by:

$$S_i = \mathbf{h}_i \, \mathbf{e} \tag{3.2}$$

$\underset{\sim}{\mathbf{h}}_i$ is a constant row vector and, on assembly of all p such rows into a full rank matrix H, the *null-space* manifold is attained as **e** reaches and remains in the intersection of p hyperplanes:

$$\text{i.e.} \quad S = \bigcap_{j=1}^{p} S_j = \{\underline{e}: H\underline{e} = 0\} \tag{3.3}$$

The subspace S is the null space of H denoted by N(H). If the N(H) manifold is reached the system phase trajectories satisfy the equation:

$$S = H\,\mathbf{e} = 0 \tag{3.4}$$

Differentiating (3.4) with respect to time:

$$\dot{S} = H\,\dot{\mathbf{e}} = 0 \tag{3.5}$$

On applying (3.5) to (3.1) and using e_{yeq} to denote the value of *equivalent control* effort which would be effective in maintaining the response within N(H), it follows that:

$$H(A_0\,\mathbf{e} - K\,e_{yeq} + E\,\mathbf{d}) = 0 \qquad (3.6)$$

If the matrix HK is non-singular, e_{yeq} can be determined from (3.6) in a unique manner as:

$$e_{yeq} = (HK)^{-1}\,H\,\{A_0\,\mathbf{e} + E\,\mathbf{d}\} \qquad (3.7)$$

Thus the *idealized* equivalent system corresponding to all motions within the N(H) subspace is given by:

$$\dot{\underline{e}} = [\,I_n - K(HK)^{-1}H\,]\;\{\,A_0\underline{e} + E\underline{d}\,\} \qquad (3.8)$$

This equation can be re-written as:

$$\dot{\underline{e}} = Deq\;\underline{e} + Eeq\;\underline{d} \qquad (3.9)$$

The disturbances influence the type of motion in two separate ways. The values of the disturbances up to the commencement of entering the invariant manifold determine the initial conditions of the equivalent system. These equations of motion are generally affected by subsequent disturbances, faults etc. The observer becomes completely insensitive to disturbances once the N(H) motion is reached, i.e. for:

$$H(I_n - K\,(HK)^{-1}\,H\,)\,E\,\mathbf{d} = 0 \qquad (3.10)$$

The reachability of N(H) depends on the behaviour of the range space R(H) motion and hence on the complete feedback design. Reachability of N(H) will be guaranteed iff:

$$\lim_{S \to 0^+} \frac{dS}{dt} \leqslant 0 \leqslant \lim_{S \to 0^-} \frac{dS}{dt} \quad \text{or equivalently if:}$$

$$\lim_{S \to 0} S\frac{dS}{dt} \leqslant 0 \tag{3.11}$$

Condition (3.11) is *necessary* for the R(H) flow to be directed towards the N(H) manifold asymptotically; it is thus a *sufficient condition* for reachability of N(H). It can be seen that for $dS/dt = S = 0$ the motion of $\mathbf{e}(t)$ lies within the N(H) manifold – thus providing the required invariancy and disturbance decoupling. **This defines the $\{H, D_{eq}\}$ - invariant subspace of the observer.**

It can be shown that the reachability conditions (3.11) are satisfied for all stable designs. Furthermore, once the N(H) manifold is reached the system is independent of **both** the disturbances and the effective control $\mathbf{e}_{yeq}$; hence in the N(H) manifold the motion of $\mathbf{e}(t)$ is rectilinear (or free) of the form:

$$\underline{e}(t) = \sum_{i=1}^{n-p} \alpha_i \underline{m}_i e^{-\lambda_i t} \tag{3.12}$$

The n–p eigenvectors $\mathbf{m}_i$, $i = 1, \ldots$ n–p form a basis set which span the N(H) subspace. The free motion thus guarantees the invariancy to disturbance terms resulting from E **d** in equation (3.1).

The n-p corresponding eigenvalues λ_i and associated right eigenvectors $\mathbf{m}_{ieq}$ of D_{eq} determine the response into and within the N(H) manifold is given by the complete observer equations (2.5) or (3.1), however, during invariancy p states can be expressed in terms of the remaining (n–p) states by substitution into equation (3.4).

The (n–p) *null-space* eigenvalues and basis vectors $\mathbf{m}_i$, for $i = 1, \ldots$, n–p satisfy:

$$H\,\mathbf{m}_i = 0, \text{ with } \mathbf{m}_i = \mathbf{m}_{ieq}. \tag{3.13}$$

The null space eigenvalues are thus the *invariant zeros* of the observer system together with the *zero state directions* $\mathbf{m}_i$, $i = 1, \ldots, n-p$. It follows then that the remaining p eigenvectors span the range space R(H). The null space eigenvectors are clearly right eigenvectors of **both** $D_0 = (A_0 - KC_0)$ and D_{eq}.

The D_{eq} *equivalent* observer system matrix defined by equations (3.8) and (3.9) has a special structure as a *singular matrix*. In order to satisfy equation (3.4), (3.5) and (3.10) it follows that:

$$H(s\,I_n - D_{eq})^{-1}\,E_{eq} = 0 \tag{3.14}$$

which can be written as:

$$\frac{1}{s}\,H\left\{I_n + \frac{1}{s}\,D_{eq} + \frac{D^2{}_{eq}}{s^2} + \ldots + \frac{D^{\ell}{}_{eq}}{s^{\ell}}\right\}E_{eq} = 0 \quad \text{for } \ell \longrightarrow \infty, \tag{3.15}$$

Equation (3.15) is satisfied as:

(a) $H\,D_{eq} = 0$, such that the rows of H are the left eigenvectors of D_{eq} corresponding to the p zero-valued eigenvalues.

(b) $H\,E_{eq}\mathbf{d} = 0$; this follows from (3.4) and (3.8).

(c) $H\,D_{eq}{}^k\,E_{eq} = 0$, for $k = 0, \ldots, n-p$ (i.e. D_{eq} is *nilpotent* with degree n–p)

It is thus clear that (n–p) modes become unobservable in the N(H) manifold, however all n modes become *uncontrollable* by $E_{eq}\mathbf{d}$ and thus by $E\,\mathbf{d}$. This corresponds to a zero-sensitivity to the disturbance signals as defined by Rozonoér

[7]. Thus the correct application of the equivalent feedback control e_{yeq} will ensure that the required invariancy conditions provided by the above analysis will be satisfied. If the matrix H can be designed to be the left eigenvectors of D_0 corresponding to the p range space modes, then e_{yeq} can be achieved through linear feedback.

A problem arises, however, as the requirement that H contains the left eigenvectors of D_0 is too restrictive as the resulting system is not asymptotically stable. The equivalent system can, however, be reached asymptotically as an example will show later in the paper.

It is considered better to work in the discrete-time domain to obviate the problem of assigning p zero-valued eigenvalues to a continuous-time closed-loop system.

Most direct digital forms of feedback system design require a discrete-time approach to analysis. It is natural then to seek to repeat the above analysis for the discrete-time case. A striking feature of the procedure is that it leads immediately to the design of a *dead-beat observer* whose properties are similar to the well known dead-beat control system [13]. Unlike the controlled system counterpart, however, the dead-beat observer provides a higher degree of insensitivity to uncertain dynamics which arises as a consequence of its intrinsic model-reference behaviour coupled to a well-determined structure of subspaces.

(c) Analysis of the dead-beat observer

A dead-beat system can only be realized in discrete-time and has no direct correspondence to a continuous-time form. The invariancy properties which result can be seen to have some similarity to the analysis given above for the continuous-time case.

The principal design technique underlying this work is that of eigenstructure assignment as this method forms the ultimately most flexible form of multivariable design tool. The eigenstructure provides the complete modal structure of any closed-loop linear system. The observer is an example of a linear system which requires disturbance-decoupling to ensure robustness. The dead-beat observer provides the required robustness by causing the response of the estimation error to reach equilibrium in minimum time.

Recent work by Frank and Wunnenbürg [14] has shown that the dead-beat observer when applied as a robust scheme for fault detection becomes one of a class of techniques based on the parity space approach of *residual generation* originally described by Chow and Willsky [15] and Lou [16] and extended recently to robust *component fault detection* [17]. The work reported here is again related to all the preceding studies with the exception that in this case, unlike earlier studies, the eigenstructure assignment approach has been adopted as the main design technique [18,19].

The discrete-time counterpart of equation (3.1) for the observer estimation error is:

$$\mathbf{e}_{k+1} = \Phi_0\, \mathbf{e}_k - K_d\, \mathbf{e}_{yk} + E_d\, \mathbf{d} \tag{3.16}$$

where $\Phi_0 = e^{A0\Delta t}$ is the matrix exponential for time step interval Δt, K_d is the discrete-time nxp observer gain matrix and E_d is the discrete-time counterpart for the disturbance distribution matrix.

By assigning p zero-valued eigenvalues to the closed-loop discrete-time matrix:

$$\Phi_{D0} = (\Phi_0 - k_d\, C) \tag{3.17}$$

We can then ensure that the matrix H is the pxn matrix of *left eigenvectors* corresponding to the zero-valued eigenvalues.

The required invariant null-space N(H) for this discrete-time case then becomes:

$$S = H\, \mathbf{e}_k = 0 \text{ (for sampling instant k)} \tag{3.18}$$

On reaching N(H) at instant k–1, it then follows that:

$$e_{k-1} = e_k = e_{k+1}, \text{ etc} \ldots$$

as the e_k etc become eigenvector directions corresponding the invariant subspace. This is clearly equivalent to the continuous-time requirement that $\dot{S}=0$ (see equation (3.5)).

It then follows that, as the N(H) manifold is reached:

$$H[\ \Phi_0 \mathbf{e} - K_d \mathbf{e}_y + E_d \mathbf{d}\ ] = 0 \tag{3.19}$$

The dead-beat behaviour of the observer will ensure that the invariant manifold is reached. If H does comprise the left eigenvectors corresponding to the zero-valued eigenvalues of Φ_{D0}, then, outside the null-space (i.e. in the range space R(H) manifold) the n–p "null-space eigenvalues" are cancelled. In the range space manifold the system has effective dimension of p with p zero-valued eigenvalues.

Once in the null-space manifold the range space eigenvalues are cancelled giving the reduced order system of order (n–p). It can thus be said that, for given starting conditions the null space will be reached in minimum time within the range space R(H) manifold. During this phase of the motion the estimation error response will only have a weak sensitivity to disturbance. The approach has constituted the basis for a very powerful robust observer design. By making the observer highly insensitive to disturbances it becomes possible to set a very low threshold for fault detection. The next section will address this problem with reference to sensor fault detection in an aircraft example.

10.4 THE SENSOR FAULT DETECTION PROBLEM

(a) Continuous-time case

When considering the diagnosis of sensor faults equations (2.4), (2.5) and (3.1) must be rewritten to include a sensor fault signal vector **f**. This is a convenient notation for representing one or more possible faults. Attention is restricted here to sensor faults, however most other forms of system malfunction can be treated in the same way. Thus in continuous-time:

$$\dot{\underline{e}} = (A_0 - KC_0)\underline{e} + E\underline{d} + K\underline{f} \tag{4.1}$$

where $\mathbf{f} \in \mathbf{R}^{px1}$; $\mathbf{Kf} \in \mathbf{R}^{nx1}$

Now $\mathbf{f} := \Delta C x$ (4.2)

which is *representative* of sensor faults. The problem is then that of designing a robust observer with a *sensor fault detection* signal S given by equation (3.4) i.e. a threshold is set on any one (or all) of the p signals given by:

$$S = He(t) \tag{4.3}$$
(fault-detect signal)

The matrix $D_0 = (A_0 - KC_0)$ is designed using *eigenstructure assignment* by assigning the required eigenvalues and eigenstructure to the *dual control* problem i.e. by assigning an eigenstructure to the matrix:

$$D_0^T = (A_0 - KC_0)^T \tag{4.4}$$

The assignment procedure is identical in both the continuous and discrete-time cases. For further details of the eigenstructure assignment algorithm the reader is referred to related work [18, 19].

If the design is carried out in the continuous-time domain then, although a dead-beat system is not achievable, an assignment can be made such that the p range space modes provide an effective robust filter. The essential requirement is to assign a null-space structure or geometry which has a complementary range space providing low sensitivity to disturbance. This can be achieved by ensuring that the range space eigenvalues are lower in magnitude than the null space eigenvalues, i.e. for:

$$|R_e \lambda_i| < |R_e \lambda_k| \tag{4.5}$$

with $i = 1, p$; $k = p+1, \ldots, n$.

The λ_i are the p range space modes of the observer.

An additional requirement is that the *fault distribution matrix* (the observer gain) K is such that:

$$HK \neq 0 \tag{4.6}$$

i.e. HK must be non-zero valued and ideally with significantly large norm. The null-space manifold will then be reachable and the large norm of the product HK will cause a significant sensitivity to the fault signal as required for rapid fault detection.

As only p estimation errors are measurable, the weighting matrix given by (3.4) must take the form:

$$H = WC_0. \tag{4.7}$$

The rows of H must satisfy equations (3.13), (3.14) and (3.15) however only one row (or linear combination of rows) need be used for the sensor fault detection weighting. The zero elements which will arise in the rows of H constrain the assignability of these rows if they are to be left eigenvectors of the closed-loop observer matrix D_0. However, the ideal structure of H can be assigned in some cases. It can be shown that the discrete-time design affords more freedom in the assignment than the continuous-time counterpart; this fact serves to strengthen the case for using the discrete-time (dead-beat) design approach.

If the rows of H *cannot* be assigned as left eigenvectors of D_0 (right eigenvectors of D_0^T) then it is sometimes helpful to exploit a degenerate eigenstructure such that the weighting can be easily determined.

The p R(H) modes are chosen to dominate the estimation error response to disturbances and faults. In practice the band-limiting property resulting from the ordering of the eigenvalues will keep the trajectory of the estimation error close to the N(H) manifold such that both S and S remain ≈ 0 even during the action of normal disturbances acting on the process. A trade-off exists, of course between the rate at which the null-space manifold is reached and the degree of disturbance insensitivity. In general by providing the highest degree of insensitivity to disturbance a very low threshold can be placed on the signals S, $\dot{S}$ or even $S\dot{S}$ for efficient detection of sensor or system (component) faults.

The lower the threshold can be placed on the fault-detect signal without the risk of false-alarm, then the more robust the monitor can be said to be.

(b) Discrete-time case

In the discrete-time case the p zero-valued eigenvalues corresponding to the range space R(H) are assigned exactly. By assigning the corresponding p eigenvectors of the *dual system* to be independent (numerically these eigenvectors form the rows of the matrix H) the dead-beat observer system is generated. The discrete-time system design allows for the discrete-time implementation and the *dead-beat* design provides the most robust possible linear realization which corresponds to the *discrete equivalent* system (similar to D_{eq} in continuous-time) can be assigned exactly.

Figure 10.1 shows a scheme of fault monitors each being driven by a different permutation of the measurements available from the plant. Each monitor (observer) must be driven by two or more measurements; in practice three is a reasonable choice, however for a low number of available measurements a choice of sets of two at a time may be preferable. The important result is that the combined operation of the monitor bank enables sensor faults to be detected rapidly and uniquely located. In the next section the procedure will be illustrated by a design example. The example has been approached by both the continuous and discrete time methods in order to illustrate a comparison and demonstrate the ultimate potential of the dead-beat observer. It must be stressed that, as only *measurable* estimation errors are processed the observers can be implemented in a very simple manner in discrete form without requiring the state-space for real-time computation.

10.5 IFD SYSTEM DESIGN EXAMPLE

The example is based on the lateral motion of a light aircraft. For this application the system is considered to be fifth order (n = 5) with three measurements available (p = 3). The linearized (or nominal) system model is used in the design and the fault monitor is tested by applying it to the fully non-linear aircraft dynamic model. The design given is for monitor 1 of the scheme shown in Figure 10.1.

Straight and level flight is not a particularly severe test for any IFD scheme, as false alarms are most likely whilst the aircraft is performing some manoeuvre (including significant parameter variations). The controller reference demands are changed during the simulated flight to model a varied range of pilotic inputs. For this study it was decided to limit the IFD scheme to the lateral motion of the aircraft as the longitudinal motion application has already been investigated [4]. The lateral motion state-space system for trim flight settings is:

$$\dot{\underline{x}}_0 = A_0 \underline{x}_0 + B\underline{u} \tag{5.1}$$

with: $x_0 \in \mathbf{R}^5$ and $\mathbf{u} \in \mathbf{R}^2$ and if $\mathbf{y} = C_0 x_0$ with $y \in \mathbf{R}^3$

$$C_0 = \begin{bmatrix} 0.0 & 1.0 & 0.0 & 0.0 & 0.0 \\ 0.0 & 0.0 & 0.0 & 1.0 & 0.0 \\ 0.0 & 0.0 & 0.0 & 0.0 & 1.0 \end{bmatrix} \tag{5.2}$$

Together with :

$$A_0 = \begin{bmatrix} -0.2770 & 0.0000 & -32.9000 & 9.8100 & 0.0000 \\ -0.1033 & -8.5250 & 3.7500 & 0.0000 & 0.0000 \\ 0.3649 & 0.0000 & -0.6390 & 0.0000 & 0.0000 \\ 0.0000 & 1.0000 & 0.0000 & 0.0000 & 0.0000 \\ 0.0000 & 0.0000 & 1.0000 & 0.0000 & 0.0000 \end{bmatrix}$$

$$B_0^T = \begin{bmatrix} -5.432 & 0.000 & -9.490 & 0.000 & 0.000 \\ 0.000 & -28.640 & 0.000 & 0.000 & 0.000 \end{bmatrix} \quad (5.3)$$

$\mathbf{x}$ = [v (side-slip vel.), p (roll rate), r (yaw rate), Φ(bank angle), Ψ (yaw angle)]^T,

and $\mathbf{u}$ = [(aileron angle), (rudder angle)]^T.

The aircraft parameter vary from the nominal model when it deviates from the specified trim conditions and this is an important consideration as any control system or IFD scheme (even the most robust) will have its performance degraded from the theoretical projection when implemented on the non-linear system.

It is thus essential to filter the effects of manoeuvre-induced parameter variations. For this lateral motion aircraft example a range-space mode of $\lambda_1 = -1.3$ has been chosen to reduce the sensitivity of the residual signal of $WC_o\,\mathbf{e}(t)$ to uncertain (unmodelled) dynamics. The null-space sub-system has been designed using four complex-conjugate eigenvalues with geometric and algebraic multiplicity one. The assigned eigenvalue spectrum is:

$$\lambda_1 = -1.3; \; \lambda_{2,3} = -17.00 \pm j\,4.13 = \lambda_{3,4} \quad (5.4)$$

The eigenstructure gives a *closed-loop* observer matrix D_0 as:

$$D_0 = \begin{bmatrix} -0.277 & 35.420 & -32.900 & -17.657 & 496.311 \\ -0.103 & -18.705 & 3.750 & -18.563 & -7.960 \\ 0.365 & -37.660 & -0.639 & -13.912 & -48.422 \\ 0.000 & 3.466 & 0.000 & -26.123 & 4.842 \\ 0.000 & -2.799 & 1.000 & -0.392 & -23.570 \end{bmatrix} \quad (5.5)$$

The normalized right eigenvectors of D_0 are:

$$\begin{bmatrix} 0.9972 \\ 0.0074 \\ 0.0741 \\ 0.0015 \\ 0.0024 \end{bmatrix}, \quad \begin{bmatrix} 0.8158 \\ 0.1798 \\ 0.2876 \\ 0.0707 \\ -0.0187 \end{bmatrix} \pm j \begin{bmatrix} 0.0000 \\ 0.1149 \\ 0.4464 \\ 0.0272 \\ 0.0291 \end{bmatrix}$$

m_1 $\qquad$ m_2, m_3, $\qquad$ m_4, m_5

For $\mathbf{h}_2 \mathbf{m}_i = WC_0 \mathbf{m}_i = 0$, for $i = 2, \ldots, 5$:

$$\begin{bmatrix} W_1 & W_2 & W_3 \end{bmatrix} \begin{bmatrix} 0 & 1 & 0 & 0 & 0 \\ 0 & 0 & 0 & 1 & 0 \\ 0 & 0 & 0 & 0 & 1 \end{bmatrix} \mathbf{m}_i = 0 \tag{5.6}$$

giving as solution:

$$W = [1 \quad -2.8786 \quad -1.2607] \tag{5.7}$$

The gains K used in the design for D_0, D_{eq} can be computed from an assignment of a generalized basis set for H.

$$H = \begin{bmatrix} 0.0383 & 0.0866 & -0.0832 & -0.0610 & 0.9900 \\ 0.0000 & 1.0000 & 0.0000 & -2.8790 & -1.2610 \\ 1.0000 & -7.7150 & 1.9860 & 0.0000 & 0.0000 \end{bmatrix} \tag{5.8}$$

$$D_{eq} = \begin{bmatrix} -4.173 & -129.019 & 39.614 & 18.655 & 0.000 \\ -0.949 & -22.403 & 2.607 & 7.289 & 0.000 \\ -1.585 & -22.063 & -9.819 & 18.923 & 0.000 \\ -0.368 & -9.751 & 1.985 & 2.367 & 0.000 \\ 0.089 & 4.496 & -2.463 & 0.377 & 0.000 \end{bmatrix} \tag{5.9}$$

The eigenvalue spectrum of Deq is:

$$\lambda_1 = \lambda_2 = \lambda_3 = 0,$$

$$\lambda_{4,5} = -17.00 \pm j\,4.3 \tag{5.10}$$

It is clear that the degeneracy assigned to D_0 results in D_{eq} having three zero eigenvalues together with one complex-conjugate pair $\lambda_{4,5}$. The corresponding complex eigenvectors $\mathbf{m}_4$, $\mathbf{m}_5$ therefore define the geometry of the null space N(H).

Note that $\mathbf{h}_2 = WC_0$ has been chosen as the actual weighting for fault detection, however W is applied as weighting to the three *measured* estimation errors [p, Φ and Ψ]. The observer system is partitioned into a {H, Deq}-invariant null-space together with a scalar robust-filter having a simple pole at $s = -1.3$. Also from (5.2) and the eigenvectors above, $\mathbf{h}_2 = WC_0$ is orthogonal to $\mathbf{m}_2$, $\mathbf{m}_3$, $\mathbf{m}_4$, $\mathbf{m}_5$

and almost orthogonal to $\mathbf{m}_1$. Hence the observability of the three weighted estimation error modes has been minimized by the eigenstructure design. Any residual sensitivity to disturbance signals will be band-limited by the s = –1.3 mode. The design allows a very low threshold to be placed on $S = WC_0\,\mathbf{e}(t)$ for fault detection. It has not been possible in this example to assign $\mathbf{h}_2$ directly as a left eigenvector of D_0.

Consequently, the scalar fault-detect signal S has the following s-domain form considering initial conditions on $\mathbf{e}(t)$, unknown disturbances and sensor faults $K\mathbf{f}$:

$$S(s) = \left[\frac{(s + 1.1)\ (s^2 + 34s + 306.1)}{(s + 1.3)\ (s^2 + 34s + 306.1)^2} \right] \{\underline{e}(0) + K\underline{f} + E\underline{d}\} \quad (5.11)$$

The reason that $\mathbf{h}_2$ cannot be assigned as a left eigenvector is due simply to the fact that the system is based on *three* measurement – yielding two zero-valued elements (n–p = 2) which limit the assignability freedom. Were it possible to assign $\mathbf{h}_2$ as a left eigenvector then n–1 modes would be cancelled by n–1 de-coupling zeros, an ideal situation which would leave only *one* mode in the residual signal.

Dead-Beat IFD Design : The dead-beat observer design is as follows. The design is based on the above example using a time step of $\Delta t = 0.01$ applied to the system defined by equations (5.1) to (5.3).

The discrete design was chosen to correspond to the continuous-time null-space eigenvalues of:

$\lambda_4 = -22.3$ ($\lambda_4 = 0.8$ in discrete-time)

$\lambda_5 = -2.0$ ($\lambda_5 = 0.98$ in discrete-time)

The range-space eigenvalues are assigned as $\{\lambda_1, \lambda_2, \lambda_3\} = \{0, 0, 0\}$. The eigenstructure gives a closed-loop observer matrix Φ_{D0} as:

$$\Phi_{D0} = \begin{bmatrix} 0.9966 & -0.7941 & -0.3274 & -0.0035 & -1.3623 \\ -0.0009 & -0.1924 & 0.0361 & -0.0010 & -0.0605 \\ 0.0036 & -5.3715 & 0.9930 & -0.0271 & -1.7218 \\ 0.0000 & -0.0010 & 0.0002 & 0.0000 & -0.0003 \\ 0.0000 & -0.0539 & 0.0100 & -0.0003 & -0.0172 \end{bmatrix} \tag{5.12}$$

The normalized *right* eigenvectors of Φ_{D0} are:

$$\underset{\mathbf{m}_1}{\begin{bmatrix} -0.6254 \\ 0.2709 \\ 0.2987 \\ 1.0000 \\ -0.6898 \end{bmatrix}}, \underset{\mathbf{m}_2}{\begin{bmatrix} 0.0230 \\ 0.0056 \\ 0.0540 \\ 1.0000 \\ -0.0020 \end{bmatrix}}, \underset{\mathbf{m}_3}{\begin{bmatrix} 0.4900 \\ 0.1812 \\ 1.0000 \\ 0.0009 \\ 0.0125 \end{bmatrix}}, \underset{\mathbf{m}_4}{\begin{bmatrix} 1.0000 \\ 0.0182 \\ 0.5343 \\ 0.0001 \\ 0.0053 \end{bmatrix}}, \underset{\mathbf{m}_5}{\begin{bmatrix} 1.0000 \\ 0.0006 \\ 0.0473 \\ 0.0000 \\ 0.0005 \end{bmatrix}} \tag{5.13}$$

For $\mathbf{h}_1 \mathbf{m}_i = WC_0 \mathbf{m}_i = 0$, for $i = 2, \dots, 5$; and using equations (5.2) and (5.6) this gives:

$$\mathbf{W} = [-0.0052 \quad 1 \quad 0.0004] \tag{5.14}$$

It should be noted, however, that $\mathbf{h}_1$, $\mathbf{h}_2$ and $\mathbf{h}_3$ are all left eigenvectors corresponding to the zero-valued eigenvalues, i.e.:

$$H = \begin{bmatrix} 0.0000 & -0.0052 & 0.0000 & 1.0000 & 0.0004 \\ -0.0002 & -0.1879 & -0.0033 & 0.0014 & 1.0000 \\ 0.0011 & 1.0000 & -0.0388 & 0.0056 & 0.2783 \end{bmatrix} \begin{matrix} \mathbf{h}_1 \\ \mathbf{h}_2 \\ \mathbf{h}_3 \end{matrix} \tag{5.15}$$

The fault-detect signal S has the following z-domain form considering initial conditions, unknown disturbances and sensor faults $K_d\,\mathbf{f}$:

$$S(z) = \mathbf{h}_1 (zI_n - \Phi_{D0})^{-1} \{ \mathbf{e}(0) + E_d\,\mathbf{d} + K_d\,\mathbf{f} \} \tag{5.16}$$

where $\mathbf{e}(0)$ is the initial estimation error vector (i.e. at time zero). If, for example:

$$\mathbf{e}(0) = \alpha_1 \mathbf{m}_4 + \alpha_2 \mathbf{m}_5 \qquad \text{for any real } \alpha_1, \alpha_2$$

i.e. the initial condition vector spans the null space N(H) then:

$$S(z) = \frac{k_0(z - 0.98)(z - 0.8)z^3}{z^3(z - 0.98)(z - 0.8)} + h_1(zI_n - \Phi_{D0})^{-1}\{E_d\underline{d} + K_d\underline{f}\} \qquad (5.17)$$

where, in this case the impulse response is completely nulled and $k_0 = 0$. This typifies the dead-beat system behaviour. This means that, given an initial condition for the estimation error $\mathbf{e}(0)$belonging to the null space N(H) only the disturbance and fault input signals have effect on the output signal S(z). It also implies that when the null-space manifold is reached the response of S(z) can be insensitive to disturbances and sensitive to certain faults depending on the design of K_d.

It can be shown that, for the general case when the initial condition vector does not belong to the null-space N(H) the initial condition response has the dead-beat behaviour.

The weighted fault-detect signal can be made very sensitive to the vector of faults **Kf** by ensuring that HK is full-rank and furthermore by ensuring that $||HK||_2$ is large. If one fault is to be detected by a particular monitor then it can be further arranged that the following inner product has a non-zero:

$$\mathbf{h}_i \mathbf{K}_j > 0 \text{ for some choice of i, j.}$$

This high conditioning or sensitivity will ensure that the jth component of the sensor fault will be detected efficiently using a low threshold. It is then clearly important to ensure, by design that the remaining products:

$$\mathbf{h}_i \mathbf{K}_k \approx 0, \text{ for } k \neq j.$$

For example, given a general $\mathbf{e}(0)$ and the requirement to detect a fault in the second sensor, the appropriate inner-product is unity (after normalization) i.e. $\mathbf{h}_1\mathbf{K}_2 = 1$. This results in the following fault-detect signal:

$$S(z) = \underline{h}_1 (zI_n - \Phi_{D0})^{-1} \{\underline{e}(0) + E_d\underline{d}\} + \frac{z^2 (z - 0.98)(z - 0.8)}{z^3 (z - 0.98)(z - 0.8)} f_2 \qquad (5.18)$$

where f_2 is the second component of the fault vector. Thus:

$$S(z) = \mathbf{h}_1 (zI_n - \Phi_{D0})^{-1} \{ \mathbf{e}(0) + E_d \mathbf{d} \} + f_2 z^{-1} \qquad (5.19)$$

which has a non-recursive, finite impulse response, or dead-beat form.

The effect of any fault components due to f_1 or f_3 have been nulled by design and, furthermore the response of the parity signal will have the characteristic dead-beat behaviour. Close to the null-space manifold the sensitivity to disturbances is minimized and hence a realistic threshold can be placed on the signal S(t) for fault detection. It should also be clear that the robust fault monitor can be implemented without the need for state-space computations as only the signal S(t), *measured* estimation errors and control signal are required.

10.6 IFD SIMULATION RESULTS

The results correspond to simulations for the continuous-time design (Figure 10.2) and the dead-beat (discrete-time) design (Figure 10.3). In both cases a step fault of 0.25 radians on the bank angle Φ has been simulated. The simulation is that of the fully non-linear aircraft including a Dryden Spectrum turbulence which acts on the longitudinal motion of the simulated aircraft. In both cases the robustness of the state estimation in the presence of non-linearity (uncertain dynamics) and turbulence is quite apparent.

The discrete-time result (Figure 10.3) shows that the null-space manifold is reached very quickly whilst the continuous-time responses of Figure 10.2 display the dominance of the range space mode at $\lambda_1 = -1.3$. It is clear from the results that the dead-beat response for the fault-detect signal exhibits a very low sensitivity to disturbance and initial conditions and consequently a lower threshold for fault detection to be tolerated. The chosen thresholds have been shown, in each case by a dotted line on the response of the *absolute* signal S(t).

Figure 10.3 shows that the dead-beat system has a complete insensitivity to the disturbance and furthermore that the estimation errors corresponding to the fault-free measurements are also insensitive to the fault signal f_2. The state estimates are thus shown to be very robust.

10.7 CONCLUSION

A model-based approach can be very effective for fault diagnosis in dynamic systems provided that the observer estimation error is made robust against modelling uncertainty and disturbances. The robust design problem is then one of disturbance-decoupling by design. In this work the original system coordinates have been retained in order to obviate the use of similarity transformations which can destroy robustness properties. The eigenstructure assignment gives a de-coupled structure such that the observer system's response has a manifold which is invariant to disturbances and parameter variations. This is the $\{H, D_{eq}\}$-invariant subspace which can be expressed in the more familiar discrete-time form as a dead-beat observer. The amount of the range space mode present in the response can be minimized by design so that the motion always remains close to the invariant subspace N(H). The dominance of the range space dynamics provides a robust filter thus reducing the sensitivity of the fault monitor to uncertain effects. This, in turn, means that a low threshold can be placed on the residual signal S, or alternatively a threshold can be placed on the signal S or its discrete equivalent. The fault monitor has thus been made insensitive to disturbances whilst sensor fault signals exceed the threshold for detection. By performing the eigenstructure assignment design in discrete-time it has been shown that the dead-beat observer can be realized very easily. The intrinsic robustness of the dead-beat system (i.e. the finite time response to setting) means that the resulting observer has a very strong degree of decoupling to unwanted disturbances, and to some extent to faults. The lowered sensitivity, however means that, in turn, a very low threshold can be placed on the "fault-detect" signal resulting in rapid and reliable fault detection even in the presence of system disturbance and uncertainty. An example has been given based on the lateral motion of a multivariable non-linear aircraft model comprising *three* measurements, *five* states, and additional uncertainties arising from *two* control inputs.

REFERENCES

1 Frank, P.M., "Fault Diagnosis in Dynamic Systems via State Estimation – A Survey", Proceedings of the 1st European Workshop on Fault Diagnosis, Reliability and Related Knowledge-Based Approaches, 2-4 Sept. 1986, Rhodes, Greece, published by Reidel Press 1987.

2 Himmelblau, D.M., "Fault Detection and Diagnosis – Today and Tomorrow", Proceedings of the IFAC workshop on Fault Detection and Safety in Chemical Plants, 28th Sept. – 1st Oct. 1986, Kyoto, Japan, pp. 95-105.

3 Patton, R.J., Frank, P.M. and Clark, R.N. (Eds), "Fault Diagnosis in Dynamic Systems: Theory and Applications", Prentice-Hall International (In Press 1988).

4 Patton, R.J., Willcox, S.W. and Winter, J.S., "A Parameter Insensitive Technique for Aircraft Sensor Fault Analysis", AIAA Journal of Guidance Control and Dynamics, July-August, pp. 359-367, 1987.

5 Luenberger, D.G., "An introduction to Observers", IEEE Transactions on Automatic Control AC-16, pp. 596-602, 1971.

6 Blackman, P.F., "Introduction to State-Variable Analysis", Macmillan Press, 1977.

7 Rozonoer, L.I., "A variational Approach to the Problem of Invariance of Automatic Control Systems", Automation and Remote Control, pp. 680-691, December 1963.

8 Wang, P.K.C., "Invariance, uncontrollability and unobservability in Dynamic Systems", IEEE Trans. on Automatic Control, AC-10, pp. 614-615, July 1965.

9 Cruz, J.B. and Perkins, H. R., "On Invariance and Sensitivity", IEEE Trans. on Automatic Control, AC-11, pp. 614-615, July 1966.

10 Wonham, W.M., "Linear Multivariable Control: A Geometric Approach", (2nd edition) Springer, New York, 1979.

11 Willems, J.C., "Almost Invariant Subspaces: An approach to high gain feedback design", PART I – IEEE Trans. on Automatic Control, AC-26 (1981), pp. 235-252, PART II – IEEE Trans. on Automatic Control, AC-27 (1982), pp. 1071-1084.

12 Mita, T., "Design of a zero-sensitive system", Int. J. Control, Vol. 24, No. 1, pp. 75-81, 1976.

13 Leden, B., "Multivariable Dead-Beat Control", Automatica, Vol. 13, pp. 185-188, 1977.

14 Frank, P.M. and Wunnenburg, J., "Robust Fault Diagnosis using Unknown Input Observer Schemes", Chapter 3 in "Fault Diagnosis in Dynamic Systems: Theory and Application", Patton, Frank and Clark (Eds), Prentice-Hall, 1988.

15 Chow, E.Y. and Willsky, A.S., "Analytical Redundacny and the Design of Robust Failure Detection Systems", IEEE Trans. on Automatic Control, AC-29, pp. 603-615, 1984.

16 Lou, Xi-Cheng, Willsky, A.S., and Verghese, G.C., "Optimally Robust Redundancy Relations for Failure Detection in Uncertain Systems", Automatica, Vol. 22, No. 3, pp. 333-344, 1986.

17 De, W. and Fang, C.Z., "Detection of Faulty components via Robust Observation", Int. J. Control, Vol. 47, No. 2, pp. 581-599, 1988.

18 Mudge, S.K. and Patton, R.J., "An Assessment of Robustness Properties of Eigenstructure Assignment in the Control of Uncertain Systems", to appear in IEE Proceedings D, Control Theory and Applications, July/August 1988.

19 Burrows, S. and Patton, R.J., "Robust Eigenstructure Assignment using the CONTROL-C Design Package", Proceedings of the 6th International Conference on Systems and Control Engineering, Lancaster (Coventry) Polytechnic, 13-15 September, 1988.

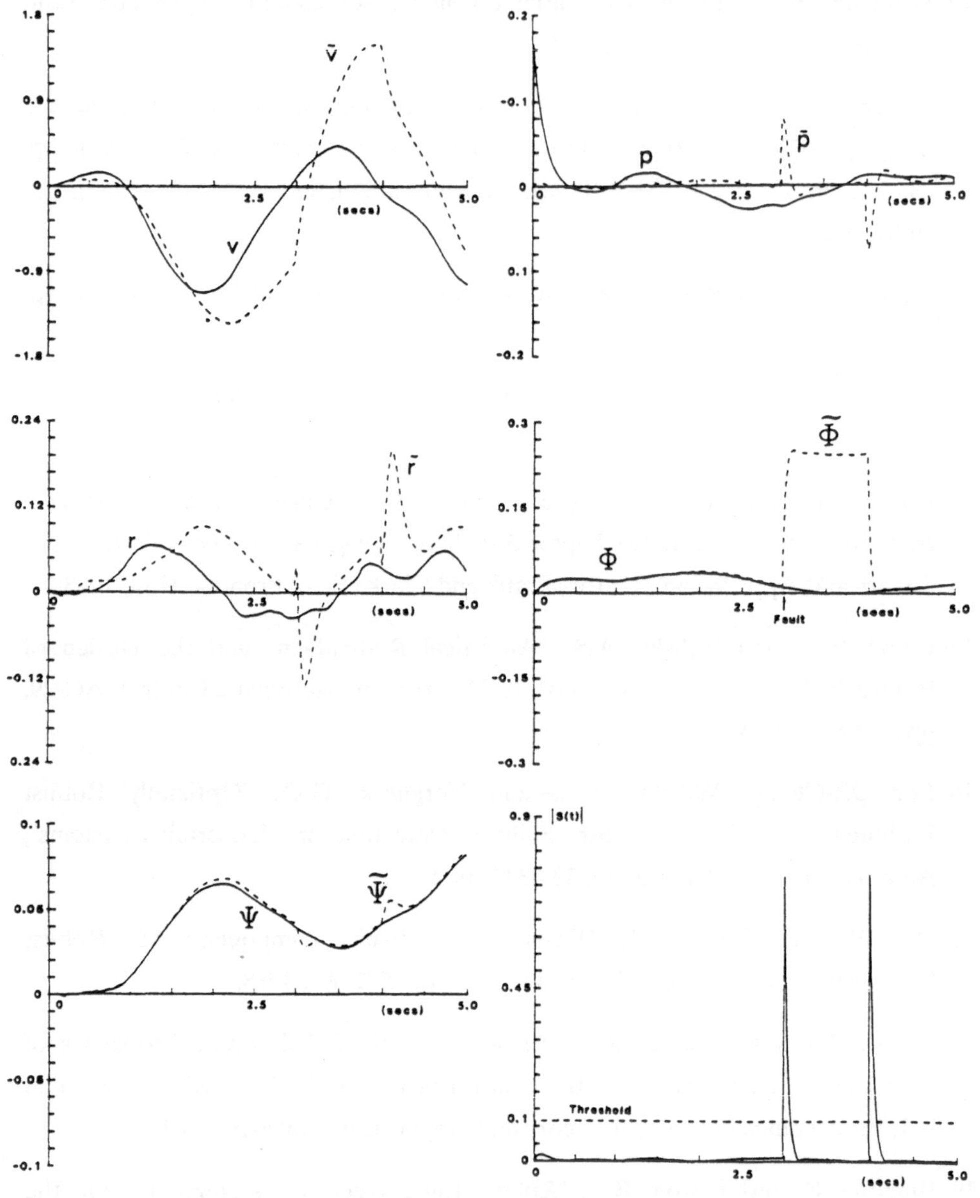

Figure 10.2. Continuous-time fault monitor responses.

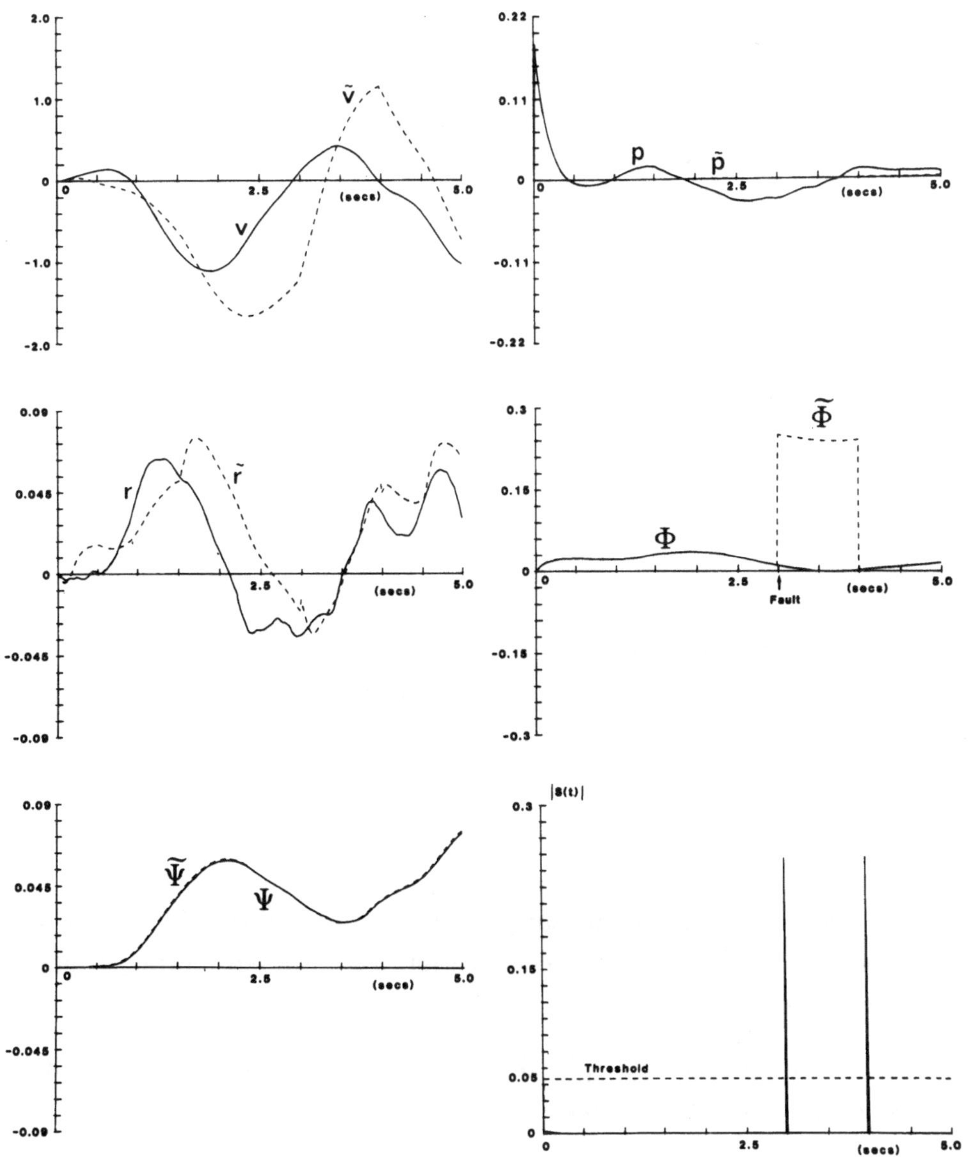

Figure 10.3. Dead-beat fault monitor response.

Chapter 11

PREDICTION OF FAILURE CONDITIONS

D. Jackson
National Engineering Laboratory
and
I. Jardine
Baker Jardine and Associates Ltd

11.1 INTRODUCTION

This chapter sets out to describe how performance simulation methods can be used in a risk assessment role and it demonstrates via a practical case study how the logistics of a safety system can be optimized. The underlying theme is to illustrate the use of a simulator as a systems design tool and how it is integrated with other reliability engineering methods. The methodology and capability are significant because they can be applied to a wide range of system optimization problems, while the actual results are of lesser interest.

Consideration is given to a pressure-surge relief system which functions as a safety protection feature of an oil terminal's exporting facility. The system in question was under construction when it was decided to analyse its functional behaviour and establish a safe operating and maintenance policy. Although under construction, minor design changes could be tolerated. Desired functional safety levels were known.

Figure 11.1 illustrates a simplified general arrangement of the system under investigation. Key features of the protection system are:

- fast-acting pressure relief valve,
- alarm systems sensing high and low pressure nitrogen,
- nitrogen supply from two sources with pressure regulators, and

- skid-mounted instrumentation package.

The proper functionality of the surge relief sub-system on demand is absolutely essential. Failure to perform correctly could lead to catastrophic failure of the loading system by overstressing and rupturing of its piping. Rupture of the piping could in turn lead to environmental pollution and present a fire hazard. Any prolonged closure of the terminal due to such events will have a significant economic impact.

A pressure surge in the export line is caused by the sudden closure of the export isolation valves downstream of the pumps. The extent of the surge depends on a number of physical factors, particularly:

- the valve characteristic – type, size, closing speed;
- the isolation valve's location in the system;
- the instantaneous exporting rate prior to isolation valve closure;
- the head of oil in the storage tanks.

The surge relief sub-system is designed to trip at a certain overpressure opening a relief valve which absorbs a slug of oil into a surge vessel and in so doing diffuses the pressure surge. The surge vessel is over-pressurized with nitrogen and having absorbed the surge will automatically exhaust the oil slug back into the loading system. In essence the surge sub-system acts like a balloon, inflated by a surge and then deflating itself after the surge has dissipated. Alarm systems will warn against too high and too low nitrogen pressure in the surge tank and in the relief valve trip device.

Because the sub-system is always under pressure it must be considered as an **active** arrangement. Its reliability must be high given the potentially catastrophic consequences if it fails to operate when a surge occurs.

By considering the surge sub-system in isolation from the export loading system an availability estimate can be predicted, that is, the fraction of its design life during which it is actually functional.

However to get a true picture of the reliability and safety of the surge sub-system it is necessary to take account of two important aspects, viz.

- The surge sub-system is not in continuous demand although it is designed to be always functional.
- There are varying degrees of consequences associated with potential failure in the sub-system.

These are very subtle and important points to appreciate which do not lend themselves to a simple analysis. Consider the following reasoning.

Because the surge sub-system is always active it can experience failures at any time regardless of whether exporting is in progress. Clearly if a failure occurs when there is no exporting taking place then the consequences of that particular failure are nil. Conversely if a failure occurs during exporting operations then the consequences are very undesirable – it would be necessary to stop exporting as a safety precaution in most cases. The foregoing implies that the consequences of faults are conditional upon exporting activities, and also assumes that all failures can be detected immediately; unfortunately this is not the case. There are a number of failure modes in the sub-system which are **covert** by nature, i.e. they cannot be detected under normal operating conditions. Covert failures are not a peculiarity of this particular system. If a serious covert fault occurs and is not detected then the protection system will not function on demand and result in a catastrophic failure of the entire system. A classic example in this case is the failure of the relief valve to open on demand.

To detect covert failures it is necessary to carry out periodic inspection and testing of equipment prone to such faults. The ability to detect and remedy covert failures in the surge sub-system is vital to its overall effectiveness.

In this case it is necessary therefore to take account of the operating and maintenance logistics as well as the hardware aspects. To analyse the problem accurately it is advisable to conduct a dynamic analysis to assess the system's reliability and logistics problems. It is very important to be able to deal with covert faults, their detection, repair and conditional consequences – this is best achieved using a simulation approach.

11.2 STUDY OBJECTIVES

The main objectives of the case study are outlined below.

(a) Establish an operating and maintenance policy capable of attaining safe working levels against catastrophic failure.

(b) Compute overall system effectiveness expressed as an availability.

(c) Compute risk of catastrophic failure expressed as allowable surge trips per annum.

(d) Check proposed spares inventory.

11.3 APPROACH TO ANALYSIS

(a) Overview

Ther are several key steps involved in reaching the final result for the overall effectiveness of the surge relief sub-system.

STEP 1 : Failure mode effects and criticality analysis (FMECA)

A failure mode effects analysis involves a rigorous assessment of each component (except piping) to establish the various modes in which the components can fail and their effect on the performance of the system. A detailed understanding of the functionality of the system is required to carry out this work. An FMECA is a **qualitative** assessment which identifies a range of faults which could arise. It can also yield suggestions for minor design modifications and safe operating procedures. An important issue in this particular study is to identify both catastrophic and covert faults.

STEP 2: Performance analysis

The next stage involves conducting a performance analysis to predict the availabilty of the system. This is a **quantitative** assessment which uses much of the FMECA results as input data. In addition, logistics aspects relating to system maintenance, spare parts, inspection requirements, available utilities etc are included. The MAROS simulator is used at this stage to enable a thorough and detailed analysis of the surge relief sub-system's overall effectiveness. Details of the simulator are discussed in Section 11.3(b). Results of this exercise yield preliminary availability estimated, IMR (Inspection maintenance repair) requirements and identify critical components.

STEP 3: Risk assessment of catastrophic failure

The final step in the analysis is to establish the likelihood of a catastrophic failure of the system. This is achieved by forming a simulation model which contains only those faults which could lead to catastrophic failure. This is known as the **risk** model which produces an availability against catastrophic failure. Using a suitable equation (discussed in Section 11.4) in conjunction with the risk model availability results and either the safety target level or the expected number of demands on the protection system it is then possible to compute the probability of a catastrophic failure.

The study was carried out in two stages; after completion of the FMECA a preliminary set of simulation runs was carried out to provide an initial feel for the availability of the system and risk of catastrophic failure. This initial set of runs concentrated on establishing inspection frequencies for detecting covert faults and identifying critical components.

With the knowledge gained from the first pass of the analysis it was then possible to define a further series of simulation runs to focus on critical and sensitive parameters. It is not wise to plan an entire set of simulation runs until preliminary results have been studied. Steps 2 and 3 were reiterated to converge on an optimun solution.

Important assumptions are highlighted in the discussion of results when appropriate, however in addition some basic assumptions are listed below.

- Revealed faults are assumed to be detected immediately.
- Equipment failures follow an exponential distribution unless otherwise specified.
- Equipment repair times follow a random pattern within a given range.
- Critical equipment repair will not be halted at the end of a normal eight-hour maintenance shift, i.e. work will continue until the repair is complete.
- All performance results are based upon 100 life cycle simulations which is considered to be more than adequate to achieve confident statistics for this particular case.

(b) Simulator background

For this exercise the MAROS simulation package was used. MAROS is an acronym for maintainability, availability, reliability and operability simulator.

MAROS function by creating typical "life cycle scenarios" of proposed systems, employing **event-driven** simulation techniques. Post-processing of these life cycle scenarios yields important information on how a system performs. By studying performance results, and altering key parameters in the design, it is possible to converge on an optimum solution.

A life cycle scenario is a chronological sequence of events which typify the behaviour of a system in real time. MAROS can create an infinite number of such scenarios for any given system, each one being unique, however sharing the commonality of being a feasible representation of how the system would behave in practice. Events are the fundamental occurrences within a system's life which determine the effectiveness of the system. The events are generated from **element data** using pseudo-random sampling techniques. Element data comprises equipment failure data, planned activities and operating logic.

Before life cycle scenarios can be created for a system it is necessary for the simulator to understand the basic system details that the user is investigating; this is achieved via a system logic model. Details of the system logic model are conveyed to MAROS by means of a spreadsheet. The user first of all sketches a logic network of the system (similar to a sideration), such as important hardware, how it functions (series, parallel, standby) and interrelates. The information is translated into a spreadsheet and supplementary details are added as required depending on the extent and objective of the performance study. Additional information may include maintenance strategies, alternative operating procedures, planned activities which interfere with the system etc. The simulator processes the spreadsheet data and sets up a digital model of the system which is acted upon by the simulation algorithm.

The simuation technique applied is commonly referred to as a "direct simulation" method. The digital system model formed from the spreadsheet data moves from one distinct state to another, governed by the occurrence of a sequence of events. The state of the model at any point in time (simulated time) is represented by a set of variables; as each new event occurs one or more of the variables representing the model changes. Progress of the simulation is in steps, from the occurrence of one event to the occurrence of the next, until the simulated time exceeds the specified design life of the system being modelled.

The simulation approach offers considerable advantages over alternative analysis methods, viz.

(a) Performance simulators conduct a dynamic analysis of a system taking account of continuous changes in the state of the system over its expected life. Account is taken of: equipment functionality; different component failure modes and consequences; concurrent/sequential/conditional events; operating and maintenance philosophies, availability of services and personnel. Clearly an attempt to calculate performance characteristics while accounting for such situations by deterministic methods (i.e. steady-state availability, fault trees, reliability block diagram techniques) becomes very unwieldy and confidence is soon lost in the ensuing results.

(b) Simulation techniques can provide a distribution of results over the life of the system, e.g. productivity profile, whereas the majority of deterministic methods provide a single expected value, e.g. time to first failure, expected number of failures etc. More insight into system behaviour can be extracted from distributed results.

(c) A distinct advantage of using performance simulators is that the user gets a true feel for the system being simulated, and learns quickly about its good points and bad points. These may relate to the design itself or its associated operations. Such insight helps the user to improve the system.

(d) Once a basic model is set up design changes can be very quickly assessed. The proverbial question "What happens if?" can be answered. Optimization of performance is a fairly rapid process.

11.4 PRESENTATION AND DISCUSSION OF RESULTS

(a) Failure mode effects and criticality analysis

The FMECA was carried out on all components given in the process and instrumentation diagrams (P&IDs) and control logic diagrams. Hard piping was not included. An FMECA is a rigorous assessment which requires a formal approach. Each item under consideration has a dedicated worksheet; the layout and contents of the completed worksheets are described below.

FMECA Worksheets: Glossary of Terms:

Comp. Code No: Each component under scrutiny is given a unique reference number of the form:

xx-yy-zzz,

where xx = circuit reference number,

yy = functional reference number, and

zzz = item number of equipment.

Failure Rate: The component failure rate in numbers of failures per million operating hours.

Repair Time: The time taken to repair the fault assuming all parts and services available, i.e. no waiting delays included.

Fm No: Failure mode number.

Fm Ratio: Failure mode ratio. The fraction of the failure rate related to the particular failure mode under consideration. The failure mode ratio is the probability expressed as a fraction that the component will fail in the identified mode. If all the failure modes of a particular component are listed the sum of the failure mode ratios for that component will equal one.

Failure Mode: The manner in which a component fails.

Effect on System: The effect of a component failure mode on the functionality of the surge system as a whole.

Severity Classification: The severity classification category assigned to each component failure mode. There are four severity classification categories:

(a) Catastrophic – Endangers personnel safety,

(b) Critical – Cannot carry out exporting operations,

(c) Marginal – Degrades exporting capability, and

(d) Minor – No effect on exporting operations.

To interface with the MAROS simulator the severity classifications are stated for the immediate effect of the fault (severity at failure) and during maintenance of the fault (severity during repair) because in many instances the consequences between these two states differ considerably and must be accounted for in the analysis.

Detection: The detection status of the fault when it occurs:

R – Revealed (detected under normal operating conditions)

C – Covert (undetected unless specifically inspected and tested).

An example of a typical worksheet is shown in Figure 11.2. Approximately 280 failure modes of the surge relief sub-system were recorded. However there are only a few of these potential faults which are of primary concern, these are:

(a) All failure modes with severity classification 1 (catastrophic)

(b) All covert failure modes of severity classification 2 (critical) at failure.

Table 11.4.1 lists those components which have catastrophic failure modes.

TABLE 11.4.1

COMPONENTS WHICH HAVE CATASTROPHIC FAILURE MODES

TAG No	Component	Failure mode	Detection
01-3-019	Pressure switch	Fails to switch	Covert
01-3-021	High pressure gas alarm	Fails to activate	Covert
01-3-036	Pressure switch	Fails to switch	Covert
01-3-039	High pressure gas alarm	Fails to activate	Covert
02-1-100	Surge relief valve	Valve fails to open	Covert
02-2-123	Rupture disc	Fails to rupture at set pressure	Covert

(b) Preliminary performance analysis

The objective of the preliminary analysis was to establish some insight into the likely performance of the system and identify critical components and operating criteria. The results of this exercise are briefly presented. Final results are reported in Section 11.4(c).

Prior to discussing these results it is important to clarify the simulation philosophy regarding equipment failure and repairs, viz.:

Revealed faults

- If a revealed fault occurs it will be entered into the maintenance log immediately. The fault will be rectified during the next or current maintenance shift according too its priority rating. The appropriate fitter will be designated to the repair work. If there is an "off-hour" maintenance crew then they will be called upon to remedy important faults outside normal maintenance hours.

- If the fault has a FMECA severity classification of 1 or 2 then any exporting operations must be stopped, i.e. the surge system is "unavailable".

- If the fault has a severity of 1 or 2 during actual repair then any exporting operations must be stopped prior to embarking on the repair work. The system becomes "unavailable".
- Given the foregoing it is implicit therefore that the system cannot incur a catastrophic failure as a result of "revealed" faults, assuming appropriate action is taken in respect of known faults.

Covert faults

- If a covert fault occurs then it cannot be detected until the next inspection period. Any faults found during inspection and testing will be remedied immediately assuming spares are available. The system is unavailable from the occurrence of the fault (which is unknown to the oeprations staff) until found and remedied during inspection. During periods of covert faults of severity classification 1, the system is in a potentially hazardous state. If a pressure surge occurs during such conditions then the system will incur a catastrophic failure.

Repair crew

- The terminal has only one mechanical fitter and one electrical fitter working on 5x8 hour shifts per week. They are not dedicated to the surge system, therefore to account for sharing resoutces with other parts of the terminal a mobilization range of 0-2 hours has been allowed. For serious faults outside normal working hours the appropriate staff can be called upon given 1-2 hours' notice.
- Operations staff are available at all times.

Overall availability

The first stage in the simulation analysis is to set up a **base case model.** This model serves as a comparison for all other derivatices. This base case model, reference "SURGE 1-1", includes approximately 200 elements, i.e. all the significant failure modes derived in the FMECA are accounted for.

The initial maintenance policy assumes that one mechanical and one electrical fitter are available 8 hours/day, 5 days/week. Adequate spare parts are assumed to be readily available and therefore will not cause repair delays. The inspection frequency is an important variable in this preliminary analysis work. The base case assumes an inspection interval of three months.

Clearly by increasing the inspection frequency the system's availability increases and more importantly the risk of covert hazardous faults is reduced. Note that inspection in this case does imply carrying out suitable testing procedures to ascertain whether a covert fault has occurred or not.

A complete model of the base case system and results is too extensive for inclusion in this paper, however an extract of some model details are given in Figure 11.3. The spreadsheet is fairly well documented and self-explanatory. Notice that each failure mode description is preceded with one of the following maintenance references:

mec – mechanical fitter required

ele – electrical fitter required

ops – operations staff required

ins – covert fault only remedied at inspection.

The preliminary set of runs concentrated on getting some insight into the sensitivity of system availability to inspection frequency of covert faults.

A range of inspection intervals from three months to one week was analysed for a 40 year design life of the surge relief system. Each life cycle simulation was repeated one hundred times to gain a representative sample of performance statistics, i.e. results are based upon 4000 years simulated operating experience. Note that each life cycle simulation creates a unique picture of how the system could behave. On this preliminary analysis all items which had convert failure modes were inspected at the defined interval – there was no discrimination between failure modes of high and low reliability items.

The results of this exercise are presented in Figure 11.4; Table 11.4.2 includes numerical availability data.

TABLE 11.4.2 SUMMARY OF PRELIMINARY RESULTS

Simulation model reference	Inspection frequency for **covert** faults	Availability mean	Availability Standard deviation
SURGE 1-1	3 months	78.5	±2.9%
SURGE 1-2	1 month	93.74	±0.7%
SURGE 1-3	2 weeks	98.61	±0.21%
SURGE 1-4	1 week	99.07	±0.13%

These preliminary results indicate that the system's availability is particularly sensitive to inspection frequency and it seems as if it will be necessary to carry out inspection and testing of critical items on a frequent basis. The most critical equipment are the alarm systems whose reliability data was cross-checked from a number of sources. Covert failures of the alarm systems accounts for approximately 75% of all losses.

From this preliminary exercise the following knowledge has been gained.

- All parties involved in the study (analyst, operator, designer etc.) now have some quantitative feel for how the system will behave. Whereas prior to this exercise no-one had any knowledge of the numbers involved.

- It is evident that the covert failure modes are an important aspect of this analysis because of the sensitivity of the results to the inspection frequency.

- Critical items have been quantified. In this case it was easy to recognize which equipment would emerge as being critical merely by viewing their reliability data.

- It is interesting to note the standard deviation results in Table 11.4.2. For the three month inspection frequency the standard deviation of the average availability is almost 3%; this relatively large variability conveys in real terms the element of "luck" in the outcome of the system over its design life — which is very undesirable. The standard deviation associated with weekly inspections is on the other hand relatively small and is indicative of "stable" behaviour. These standard deviation results produced from the simulation data are a useful parameter in conveying performance fluctuations and are used to define confidence levels.

The first pass of the analysis has shown some trends and provides a useful basis for further discussions with designer and operator personnel. However an important question at this stage is "What level of availability is required *vis-à-vis* minimum safety requirements?" This point is elaborated in the following section.

Risk assessment

To compute the risk of catastrophic failures a simulation model containing only the equipment failure modes of severity classification 1 (i.e. catastrophic) was analysed. This is known as **risk** model. The system will have a higher availability with respect to catastrophic failure than to the set of all failures established in the previous section. The catastrophic failure model, reference filename "SURGE 2-1", assumes an inspection frequency of one week. The results are as follows, quoted as mean and standard deviation.

Availability against catastrophic failure : 99.86 ± 0.05 %

Equivalent overall availability: 99.07 ± 0.13 %

This information is a prerequisite at this stage to get some feel for the safety of the system in line with the minimum desired safety level.

To compute the risk of catastrophic failure the following formula is used.

$$Pcf = 1 - (Pd^{**}X),$$

where Pcf = Probability of catastrophic failure over working life,

Pd = The probability of the system being available to cope with catastrophic failure modes = availability/100,

X = Total expected surges in life of system, and

** = Raised to the power.

If the safety level against catastrophic failure is known along with the expected number of surges per annum then using the above formula will yield a minimum availability target against catastrophic failure, viz.

Taking a defined minimum safety level against catastrophic failure as 0.001/annum and an expected number of surges per annum as 1, the availability target is:

$$0.0001 = 1 - Pd$$

$$Pd = 99.90\%.$$

Comparing the minimum target (99.90%) with the "best" preliminary results (99.86%) indicates that the system will at present not meet minimum safety standards. It is necessary to review the design, operations and maintenance aspects to improve overall effectiveness. This review should be conducted along with the appropriate staff involved in the project. From a practical viewpoint weekly inspection/testing seems to be too frequent for this particular situation. A longer period between inspection would be desirable while attaining required safety levels.

(c) Final analysis

The high pressure alarms were found to be the most critical items and they incurred potentially catastrophic covert faults. A design change for all the alarm systems was advised which introduced alarm test features directly into the main control system. All alarms would be tested (by push-button) prior to any exporting session. The effect of this design modification was twofold: the burden of frequent inspection of the alarms by the maintenance crew was alleviated, and the potential for covert faults lying dormant for long periods was considerably reduced. Note however the covert alarm faults were not eliminated. Their maximun exposure time was minimized to the length of one export session (less than one day) compared with the period between inspections in the preliminary analysis. On the negative side a few more failure modes were introduced because of the extra electrical equipment, and these would tend to reduce overall reliability.

Preventive maintenance strategies were to be investigated for critical items which affect the safety of the system. A schedule of planned replacement before "expected" failure was assessed using the simulator. Only equipment which exhibits "wear-out" should be considered for preventive maintenance and the problem then is to establish an optimum threshold before imminent failure. Equipment failures described by the exponential distribution are not eligible for preventive maintenance because this distribution implies that newly replaced equipment has the same failure rate as the items replaced and will therefore yield no gains. In this study candidates for preventive maintenance were described by a Weibull distribution with shape factor greater than unity. Raw failure data, including censored data, can be fed into the simulator to produce a Weibull data fit. The threshold for planned replacement is described as a ratio of the characteristic life of the equipment concerned. If a valve has a characteristic life of, say, 30 years and its replacement threshold is specified as 50% (of the characteristic life) then it will be replaced after 15 years service – the valve has still a potential to fail prior to its threshold.

Spare parts were also to be reviewed to check that the proposed inventory was adequate. When modelling spares there are three importan parameters:

- The number of each "type" of parts in stock
- The lead time required to replenish the stock
- The time required to mobilise the parts to the repair site.

Optimization of risk model

The first step in this work is to establish a new base case model to serve as a yardstick for the optimization procedure. This new base case model, reference SURGE 12-1, is similar to the previous risk model (SURGE 2-1) but the design, the operating and maintenance policies have been changed to reflect the new issues discussed above.

The preliminary analysis assumed that inspection of all critical items was carried out at the same frequency – this simplification did not account for the wide range in their reliability. In this optimization account will be taken of the relative reliability of the critical items in an attempt to reach a sensible balance between inspection frequency and item reliability.

Since assuming the introduction of test buttons on the alarms, the remaining critical items which could lead to catastrophic failure are:

- the pressure relief valve
- the pressure switches
- the surge vessel rupture disc.

Note that the rupture disc failure is a "second order fault", i.e. to cause catastrophic failure there needs to be an abnormally high surge beyond the design pressure of the surge vessel **and** the disc needs to fail to rupture. Faults in the rupture disc which would prevent it from rupturing cannot be detected by inspection. As a result of the foregoing the rupture disc faults cannot be realistically assessed and will not be considered further.

The first series of simulations investigated varying the inspection frequencies of the critical items. Results of these runs are tabulated below.

TABLE 11.4.3

AVAILABILITY AGAINST CATASTROPHIC FAILURE

Simulation model reference	Inspection frequency for **covert** faults	Availability Mean	Availability Standard deviation
SURGE 12-1	Relief valve annually Pressure switches monthly	99.70	±0.55%
SURGE 12-2	Relief valve at 6 months Pressure switches monthly	99.75	±0.33%
SURGE 12-3	Relief valve at 6 months Pressure switches monthly	99.91	±0.21%
SURGE 1-4	Best preliminary result for comparison	99.86	±0.05%

By comparing these new estimates with the best results in the preliminary analysis it is still evident that frequent inspection is required. Two points should be noted when making these comparisons.

- A 0.1% change in results is significant in this case when catastrophic failure is being considered.
- The standard deviation of the latest results is significant and concerning. This relatively high standard deviation reflects that with the latest inspection strategy there is a higher chance of not being aware of a serious fault.

To put the problem in perspective the number of potentially serious faults, from the relief valve and pressure switches, in 40 years is only around one! Inspection may not seem to be an important issue therefore, however this statistic also reveals that if inspection of covert faults was not carried out at all then the risk of catastrophic failure becomes almost a certainty!

The next series of simulations concentrated on assessing the benefit of planned replacements of the critical items prior to their failure in an effort to offset any serious faults. This preventive maintenance philosophy also included assessing the replacement of the high-pressure gas alarms.

To model preventive maintenance and assess any possible benefits it is necessary to establish a wear-out characteristic for the equipment concerned. Note that all previous analysis assumed that equipment failures were based upon a constant failure rate, i.e. they conformed to exponential failure theory, which is the most common method of assessing reliability, primarily because of its simplicity. Exponential theory states that new equipment is just as likely to fail as old equipment and is therefore unsuitable for preventive maintenance analysis. Consequently all items considered for planned replacement were modelled using Weibull theory which accounts for either burn-in or wear-out. A wear-out shape factor of 3.0 was used in all cases.

Results of this exercise are very encouraging; refer to Table 11.4.4. Unless otherwise specified the inspection interval for the relief valve was six months and one month for the pressure switches.

TABLE 11.4.4

EFFECT OF PREVENTIVE MAINTENANCE

Simulation model reference	Replacement Strategy	Availability Mean	Availability Standard deviation
SURGE 13-1	Relief valve at 10 yrs Pressure switches at 5 yrs HP gas alarms at 1 yr	99.95	±0.002%
SURGE 13-2	Relief valve at 15 yrs Pressure switches at 7.5 yrs HP gas alarms at 1yr	99.96	±0.002%
SURGE 13-3	Relief valve at 20 yrs Pressure switches at 10 yrs HP gas alarms at 1 yr	99.96	±0.012%
SURGE 13-4	Relief valve at 20 yrs Pressure switches at 10 yrs No gas alarms replaced	99.96	±0.012%
SURGE 13-5	Relief valve at 20 yrs Pressure switches at 10 yrs HP gas alarms at 1 yr Switches now inspected at six month intervals	99.95	±0.115%

The results show a definite advantage in planned replacement of critical items. Replacement of the relief valve could be construed as a major overhaul or replacement of internals only. SURGE 13-5 also indicates that inspection of the pressure switches should be around one month intervals as opposed to half yearly because of the increased standard deviation with the latter. Replacement of the HP gas alarms does not yield improved performance. All these results exceed the minimum safety level of 99.90%.

To investigate the stability of these predictions a sensitivity analysis was carried out. The SURGE 13-2 model was taken to represent an optimum and its component reliabilities were all modified by ±25%.

Table 11.4.5 contains the results.

TABLE 11.4.5

SENSITIVITY ANALYSIS RESULTS

Simulation model reference	Reliablity data variation	Availability	
		Mean	Standard deviation
SURGE 13-2	Basic	99.96	±0.002%
SURGE 14-1	−25%	99.95	±0.022%
SURGE 14-2	+25%	99.96	±0.002%

The results are very stable, indicating that the outcome of this analysis is not sensitive to reliability data input (which is generally accepted as being the most concerning parameter). The performance of this system is dominated by its operating and maintenance policies. A good degree of confidence should be attached to these predictions.

Having arrived at an optimum availability against catastrophic failure, the risk of catastrophic failure can now be established as per Section 11.4(b) (Risk assessment). However now that a design target has been stipulated (0.1%/ annum) it is now possible to define the number of surges per annum which the system can cope with within its acceptable threshold, viz:

$$\text{Pcf} = 1 - (\text{Pd} ** \text{X}),$$

hence $$\text{X} = \log(1 - \text{Pcf}) / \log(\text{Pd}),$$

where Pcf = Design target /annum,

Pd = Probability of the system being available to cope with catastrophic failure modes = availability /100, and

X = Acceptable number of surges per annum,

hence

$$\text{X} = \log(1 - 0.001) / \log(0.9996),$$

where X = 2.5.

Concluding, given a design target of 0.001/annum against catastrophic failure, the system as described by the optimum model SURGE 13-2 (reference Table 11.4.4) can cope with up to 2.5 surges per annum. The expected number of demands on the surge relief system is around one per annum, hence the safety factor for the system is around 2.5 above mimimum requirements.

Overall system availability

To complete the analysis an overall system availability was computed based upon the optimum risk model. A check was made to see if the spares inventory was adequate at this stage by comparing results with and without the inventory simulated.

Basis for the final analysis:

- Design life : 40 years.
- Maintenance crew on 5 * 8 hour day shifts per week, 0-2 hours to mobilize. Backup crew available for critical faults beyond normal maintenance hours at 1-2 hours call-out.
- Spares inventory modelled.
- Inspection of covert faults (which imples conducting necessary testing) as follows:

    (a) relief valve at six monthly intervals

    (b) pressure switches at monthly intervals.
- All alarms have test buttons fitted and these are tested prior to each lading session.
- The relief valve will be replaced (or given major overhaul) after 15 years' service.
- Pressure switches to be replaced at 7.5 year intervals.

Final overall system availability predictions expressed as mean and standard deviation are:

overall system availability – 99.29% ± 0.65%

effect of spares inventory – -0.01%.

The effect of the spare parts policy is to reduce the overall affectiveness by only 0.01% which is an almost negligible amount, indicating that the policy is suitable.

11.5 CONCLUSIONS

Simulation, by creating typical life cycle scenarios of proposed systems, allows key design parameters in the design to be changed and makes it possible to converge on an optimum solution.

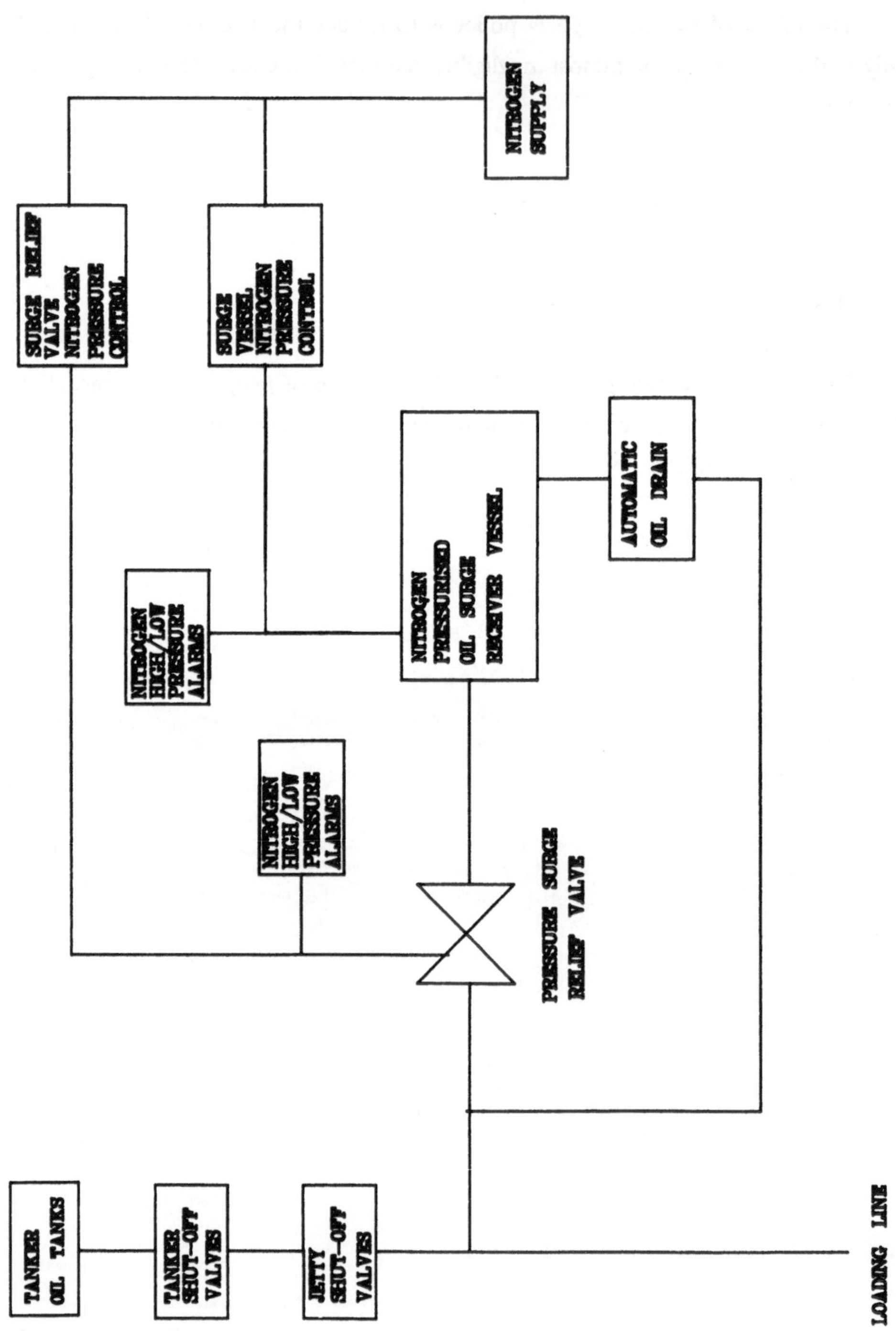

Figure 11.1. Simplified general arrangement of a pressure surge relief system.

Department of Trade and Industry
NATIONAL ENGINEERING LABORATORY
East Kilbride Glasgow G75 0QU

FAILURE MODE EFFECTS AND CRITICALITY ANALYSIS
Failure Mode Effects Analysis

Customer: Subject: SURGE RELIEF SYSTEM.   Sub-system: OIL SURGE RELIEVER/RECEIVER.

Component code no: 02-1-104 Description: SHUT-OFF VALVE:GATE TYPE: MANUAL OPERATION.

Failure rate: 2.00 Operating time: 40 YEARS. Repair time range hrs: 2-5. Sheet 5 of 30

FM NO	FM RAT	FAILURE MODE	EFFECT ON SYSTEM	SEVERITY@FAIL	SEVERITY@REPAIR	DETECTION
01	0.12	VALVE FAILS TO OPEN.	LOSS OF SURGE VESSEL HIGH OIL LEVEL ALARM. SURGE VESSEL HIGH GAS PRESSURE ALARM WOULD OPERATE.	2	2	C
02	0.12	VALVE FAILS TO CLOSE.	OPERATIONALLY NONE. DELAY IN MAINTENANCE OPERATIONS.	4	2	R
03	0.75	LEAK TO LOW PRESSURE OR ATMOSPHERE.	NONE. SITE NITROGEN SUPPLY WOULD COPE WITH MINOR LEAK.	4	2	C
04	0.01	MAJOR LEAK.	SURGE VESSEL LOW PRESSURE ALARM OPERATES.	2	2	R

Figure 11.2. Example of a typical FMEA worksheet.

```
*************************************************************************
*                                                                       *
*                  SYSTEM PERFORMANCE SIMULATOR                         *
*        copyright Baker Jardine and Associates Ltd.  1988              *
*                                                                       *
*************************************************************************
| SPREADSHEET for MAROS release 3.1
|------------------------------------------------------------------------
|
| JOB ID  : OIL TERMINAL SURGE VESSEL RISK MODEL
|
| REMARKS :        RISK MODEL against Catastrophic failure
|
|                  with Opportune replacement of Critical Items
|
|                  a. Pressure Relief valve every 15 years
|
|                  b. Pressure switches every 7.5 years
|
|                  c. HP Gas Alarms every year
|
|    NOTES :
|
|    1. Failures have been split into two categories, viz:
|
|       a. Revealed failures which are detected immediately.
|
|       b. Covert failures which can only be detected by inspection
|          and possibly need for testing.
|
|    2. Effect of failures is split into two categories, viz:
|
|       a. Immediate impact of failure.
|
|       b. Impact during repair.
|
|
|    3. Repair of 'covert' failures can only be carried out during
|       inspection periods.
|
|       The following inspection sequences have been set-up, which
|       differ from the preliminary analysis :
|
|       a. Pressure relief valve checked every 6 months
|
|       b. Pressure switches checked once per month
|
|
|    4. Maintenance services assume day shift cover (8 hrs x 5 days)
|       by Mechanical and Electrical Fitters with mobilisation 0-2 hrs.
|       to allow for sharing of services with other systems on STURE.
|       There is now a backup 'off hours' capability with mobilisation
|       1-2 hours anytime outwith normal hours.
|
|    5. Spares are not modelled in this case because they are not having
|       any effect on the risk results.
|
|    6. System design life is 40 years
|
|    7. This model assumes all Alarm Systems are now fitted with a test
|       button which will be checked prior to each loading sequence.
|
|------------------------------------------------------------------------
| SETTING-UP PARAMETERS
| SYSTEM LIFE     :  40.     years.       NUMBER of SIMULATIONS  : 100
| KEYWORDS        :  annual
|------------------------------------------------------------------------
| ELEMENT DATA (Definition of System Model Contents)
| Nr. of ELEMENTS :          [max. 1000]     LIFE CYCLE EVENT LIMIT:
| Part 1  UNSCHEDULED EVENTS (Component failures) :-
| ID |                          |  FAILURE    |   REPAIR   |CAPAC. LOSS|BRNCH
| Nr.|    DESCRIPTION           |   DATA      |    DATA    |@FAIL,@REPR| Nr.
|....|..........................|.............|............|...........|.....
C
C
C   LOGIC/CONTROL SYSTEM
C   --------------------
C ---comp. code : 01-3-019
  1 |in2:PRESSURE SWITCH fm01  ,   81.3      |  .05,.1,-1  |   100     |1
C ---comp. code : 01-2-036
  2 |in2:PRESSURE SWITCH fm01  ,   81.3      |  05,.1,-1   |   100     |1
C ---comp. code : 02-3-021
  3 |in1:HP GAS ALARM fm01         2 7.3     |  05,.1,-1   |   100     |1
C ---comp. code : 01-2-039
  4 |in1:HP GAS ALARM fm01         2 7 3     .  05,.1,-1   |   100     |1
```

```
C
C
C   SURGE VESSEL DETAILS    -  Oil Surge Receiver/Reliever
C   --------------------
C
C ---comp. code : 02-1-100
  5 |in3:RELIEF VALVE fm01      |   230,3     | .3,.5,-1  |   100     |1
C ---comp. code : 02-2-123
  6 |mec:RUPTURE DISC fm01/02   |    114      | .1,.3,-1  |   100     |1
C
.............................................................................
 Part 2  SCHEDULED EVENTS (Planned interventions) :-
 ID |                        |FREQ  | START | END   | ACTIVITY |CAPAC.|BRNCH
 Nr.|      DESCRIPTION       |(yrs) |yymmdd |yymmdd | duration | LOSS | Nr.
....|........................|......|.......|.......|..........|......|.....
 10 |in2:Renew press switches| 7.5  |080601 |       |      .1  | 100  |  1
 11 |in3:Renew Press Rel Valve| 15  |160101 |       | .3,.5,-1 | 100  |  1
 12 |in1:Renew HP alarms     | 1    |020101 |       |      .1  | 100  |  1
.............................................................................
 Part 3  CONDITIONAL EVENTS (Operability logic)
 ID |                        |  FAILURE    |   REPAIR    |CAPAC. LOSS|BRNCH
 Nr.| Describe ACTIONS/EVENTS|   DATA      |    DATA     |@FAIL,@REPR| Nr.
....|........................|.............|.............|...........|.....
    |                        |             |             |           |
.............................................................................
 NETWORK DATA
 Nr. of BRANCHES :            [max. 100]
BRANCH|BRANCH|
  Nr. | CAP. | BRANCH UPSTREAM CONNECTIONS  [use list format]
......|......|...............................................................
   1  | 100  |
-----------------------------------------------------------------------------
 SERVICE UTILITY DATA  [vessels, manpower, tools, spares]
 Nr. of UTILITY GROUPS :  5        [max. 100]
CODE|       DESCRIPTION      | Nr. |MOBILISATION |AVAILABILITY | span |
 Nr.|                        | off |range (days) |   DATA      |      |  freq.
....|........................|.....|.............|.............|......|.......
 1  | Mech. Fitter           |  1  |  0.0 ,0.08  | 8 hrs/day   | 5,7  |
 2  | Elect. Fitter          |  1  |  0.0 ,0.08  | 8 hrs/day   | 5,7  |
 3  | Off hours crew,m       |  1  |  0.04,0.08  |             |      |
 4  | Inspection crew,m      |  2  |             | 0201,0203   |      | 0.0833
 5  | RelVal Insp Crew,m     |  1  |             | 0601,0602   |      | 0.5
-----------------------------------------------------------------------------
 WORK ZONE & EQUIPMENT LOCATION DATA
ZONE|     EVENT LIST  [enter list of maintainable equipment per location ]
 10 |                 [eg. 5-10,18,65-70  repeat ID on continuation lines]
....|........................................................................
C
C   All equipment on the same work zone
C
 1  |   1 - 20
-----------------------------------------------------------------------------
 JOB PRIORITY DATA
 REMARKS :  Priorities 1 - 5   -> level 1   NB : refer to users guide for
            Priorities 6 - 99  -> level 2        'level 1-2' logic
 Priority | Keyword | Event Nr              | Capacity Loss at Repair
          |(6 chars)| accept list           | single value or range
..........|.........|.......................|................................
C --default all work to priority 5
    5     |         |  1 - 20               |  0 - 100
C --top priority to any fault causing loss of functionality of surge system
    1     |         |  1 - 20               |  100
-----------------------------------------------------------------------------
 MAINTENANCE UTILITY CROSS REFERENCE DATA
 Prime|Backup| Extra| Event Nr              | Keyword | Work Zone ID's
 Group| Group| Group| accept list           |(6 chars)| accept datalist
......|......|......|.......................|.........|......................
   1  |  3   |      |                       |  mec    |
   2  |  3   |      |                       |  ele    |
C ---- Alarms tested on a daily basis and maintained by electrical fitters
   2  |  3   |      |                       |  in1    |
   4  |      |      |                       |  in2    |
   5  |      |      |                       |  in3    |
-----------------------------------------------------------------------------
 PLANNED MAINTENANCE DATA
 Event ID |   List of Equipment to be RENEWED (reference via event ID)
..........|..................................................................
   10     |     1,2
   11     |     5
   12     |     3,4
.............................................................................
```

Figure 11.3. Extract from simulation model.

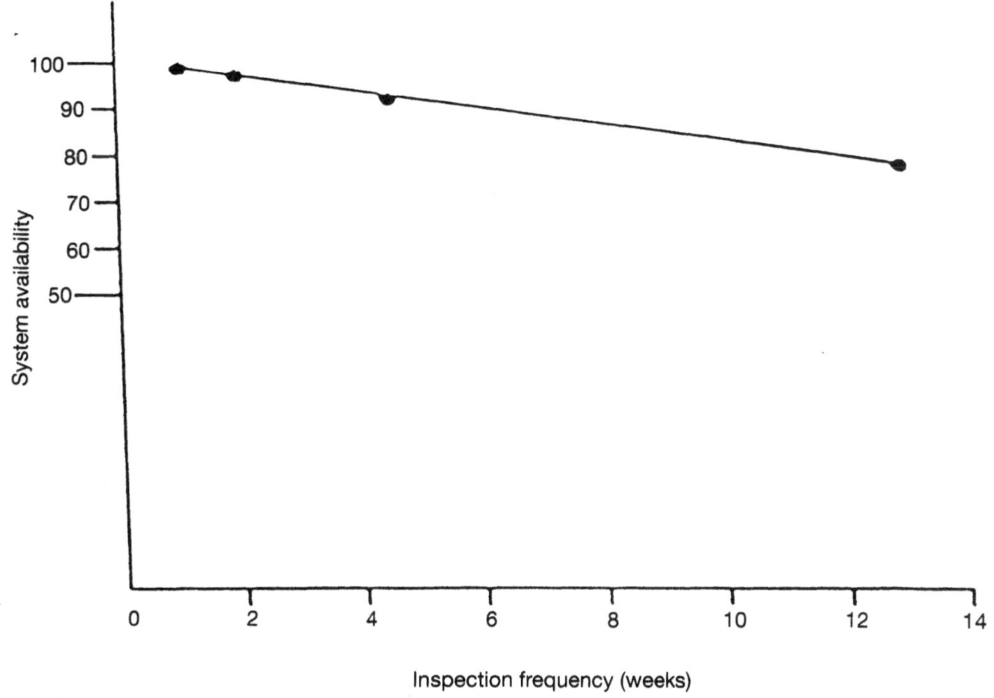

Figure 11.4. Sensitivity to inspection frequency.

Chapter 12

ACCIDENT MANAGEMENT AND FAILURE ANALYSIS

G. C. Meggitt
Safety & Reliability Directorate

12.1 INTRODUCTION

In the design of systems with safety implications the initial aim is to make them totally fail-safe. This can sometimes be achieved but in complex systems there may remain some conditions where fail-safe cannot be guaranteed. If these conditions are predicted to occur with sufficiently low frequency then the system design may be considered acceptable, but it may still be desirable to make provision for mitigating the consequences of the events. In this presentation some of the factors which are involved in emergency management are discussed. The general principles developed are illustrated using examples drawn from nuclear emergency response management.

12.2 NUCLEAR SAFETY

In order to obtain approval to construct a nuclear reactor or chemical plant in this country it is necessary to meet the stringent design standards set by the Health and Safety Executive. The standards require, among many other things, that failures of the plant which would result in radiation exposures of members of the public should be possible only with acceptably low frequencies. This is achieved by employing fail-safe technology, redundancy, diversity and defence-in-depth; its achievement is demonstrated by the application of the range of techniques of reliability and safety assessment. Of course, it is necessary to show that the plant will be operated — as well as designed — to acceptable standards of safety.

12.3 THE ACCIDENT

An accident is, when it happens, unexpected. The theoretical possibility of an accident may be well-established but each one is an unpleasant surprise. Since we manage our affairs in a way which deals with expected events most efficiently, an accident is likely to need a restructuring not only of data acquisition and assessment systems but also of the actual way we think. Priorities are changed and we may be faced with handling situations of which we have no prior experience.

Such situations are suffused with stress, particularly if there is pressure to take action quickly to save plant and people. They may be complex and require analysis to identify the correct course of action for retrieval. The combination of these factors may lead to inefficient and even counterproductive action if basic preplanning has not been undertaken.

12.4 THE ACCIDENT RESPONSE

For the purposes of this discussion two types of response to accidents will be distinguished: the automatic and the tailored. These are not truly distinct but separating them will point up important factors.

12.5 THE AUTOMATIC RESPONSE

In the nuclear industry there is one situation in which automatic response is particularly important: the criticality accident.

Criticality results from the coming together of more than a critical mass of fissile material such as U-235 or Pu-239. It is the essence of the industry; nuclear reactors with a critical mass under carefully designed conditions and electricity is produced by the energy resulting from the chain reaction of the fuel in the critical arrangement. The hazard comes from the possibility of the accidental creation of a critical mass in a fuel production, reprocessing plant or experimental facility. Should this happen the result would be the generation, possibly extremely quickly, of large quantities of energy and very high radiation levels close to the critical mass. The levels would be so high that exposure to them for even a few seconds could be fatal.

The response to this possibility is first of all to make it as remote as possible through a system of physical and administrative controls which make it extremely unlikely that a critical mass could be assembled accidentally (we have not had a critical accident in the UK which caused radiation exposures of workers). However, even when this is done, detectors are installed which could respond to the high radiation fields very quickly. These detectors sound alarms and the instructions to workers are to evacuate the area as rapidly as possible. This rapid evacuation is required because of the need for workers to distance themselves quickly from the location of the accident (the inverse square law operates); a quick response could make the difference between a non-fatal radiation dose and a fatal one. Such an evacuation which carries with it risk of plant damage, is traumatic and could result in personal injury. It is therefore worth looking at the characteristics of the instrument on which so much reliance is placed: a device which will detect an accidental criticality and yet be virtually proof against false alarms.

For reliability the detectors are as simple as possible and use well-tested components. To avoid false alarms they are generally operated in voting arrangements (say two out of three) and they detect the rate of rise of radiation fields. The combination of these factors means that they are unlikely to be triggered by radiation sources normally present in an area when the sources are moved around. A further feature contributing to the reliability is the noise familiar to those who have seen television programmes about the nuclear industry – the regular chirping of the test signal. This is fed into the system to check that it is working properly.

This example illustrates a number of factors concerned with automatic responses to accidents:

- It must be possible to identify the incident when it happens. This means that it should not be confused with other incidents (which may demand different actions) and there should not be too many false alarms. False alarms are not only expensive and disruptive but their frequent occurrence degrades the response.
- It must be possible to specify a course of action following the accident which is beneficial, if not in every aspect, at least overall. In the case of the criticality accident it has been decided that personnel should not attempt to evaluate the situation or even to pause to locate the accident so that they can run directly away from it. Instead, they are instructed to run along a predetermined route which takes them out of the area. The route is planned balancing the need to get out quickly against the possibility of a worker actually running past the accident location.
- The warning of an accident must be given in unambiguous terms and the people involved must be absolutely clear on what they must do. The first requires that the accident warning be clear and impossible to confuse with some other signal. The second implies that the action to be taken should be as simple as possible and should preferably, be rehearsed regularly.

The instructions for action in a criticality accident require people to report to a centre, remote from the incident location. Here an assessment is made of the radiation doses received by personnel and this forms one of the bases for the tailored response which will involve necessary medical treatment and decontamination and evaluation of the accident.

12.6 THE TAILORED RESPONSE

The response to many accidents passes through an automatic phase but quickly reaches a stage where a response must be adapted to the actual situation. The demands of this phase are at least as difficult to meet.

Several functions of the accident management system can be identified:

- Acquire relevant information.

- Assess the information.
- Act on the information.
- Inform others.

These may sound straightforward but it has to be recognized that we are discussing rapid response to an unexpected situation. Systems for the acquisition and processing of data are probably optimized for normal conditions; in an accident a completely different set of data must be collected, evaluated and acted upon quickly and with, perhaps, minimal warning.

As an example here we will consider a management system planned to deal with a nuclear accident which might affect the public.

The accidents at nuclear establishments to which greatest attention has been devoted are those which result in release of radioactive material to the atmosphere. This radioactivity would drift on the wind and disperse and dilute it as it travelled and, if inhaled, would irradiate organs of the body. Some materials like the noble gases, are not incorporated into the body in this way, but nonetheless may give rise to radiation exposures as the cloud of radioactivity passes people. Some of the released material may be deposited on the ground; if it is, it leads to radiation exposures over a longer period and it may also be taken up by foodstuffs and eventually consumed by humans.

The consequences of radiation exposures may be acute; if the radiation doses are very high people will become ill and perhaps die within a few weeks. Lower doses will have no immediate observable effect but they may increase the risk of dying from forms of cancer later in life.

(a) Acquisition of information

The possibility of a release to the atmosphere may be indicated first by a fault on the plant. Generally such faults will have been identified by the hazard assessment of the plant and there will be back-up systems and filtration to mitigate the consequences; plant instrumentation may give estimates of the quantity and type of radioactive material involved. The final plant instrumentation before a release occurs will be the monitor fitted to the discharge stack.

The stack will be instrumented to record the small quantities which may normally be discharged. In an accident, it may receive a much greater quantity and it would therefore be good practice to ensure where practicable that the dynamic range of the device is sufficient for it to measure these greater discharges. A fault analysis should assist in this judgement.

The stack monitor should therefore be one source of information on the public impact of the accident. There are, however, two other sources of data.

In recent years many nuclear facilities have had radiation monitors installed around their perimeters with provision for telemetry of the results to a central station. These devices can give a continuous record of radiation levels in real-time and, since they are are run continuously, are sure to detect whatever releases of radioactive material have occurred – provided of course, that they are sufficiently sensitive.

In addition to these truly real-time monitors it would be normal practice, once a release is suspected, to send out survey teams in vehicles to collect the passive radiation monitors present around all nuclear sites. They would also take measurements themselves. It is possible by this means to make a wider range of measurements: airborne activity, radiation from the ground and activity deposited on the ground and plant material.

The installed monitors constitute an essentially automatic recording system. The survey teams must, however, be directed in order to obtain useful information.

When an emergency is declared on a nuclear site, several teams of people will assemble (indeed this could be classified as an automatic response). The team dedicated to collecting information from surveys is, perhaps, the most important in the very early phases. It is responsible for directing the survey teams to the correct areas to obtain relevant data.

As part of the emergency plan a number of roads which encircle the site will have been identified and particular monitoring points designated. The survey teams would be briefed before being sent to collect the passive monitors downwind of the affected plant, make standard measurements and collect predetermined samples of vegetation. The results of the measurements would be reported by radio as soon as they were taken. It has been found helpful to do this in a standard form to ensure that errors in transmission are minimized and, almost as important, that the complete data can be retrieved later.

The control of the survey teams is in the hands of an experienced health physicist in direct contact with the radio operator. Before him he will have a detailed map of the area and the results from the survey teams will be displayed on the map. He will concentrate on establishing the levels of radioactive contamination in the environment around the plant and advising his survey teams on necessary personal protective measures. In the initial stage it is most likely that there will be a relatively small number of spot measurements from the survey teams.

In summary, the information in the early stages of the accident – ignoring any which may be derived from the plant itself – is likely to be:

- The discharge measured by the stack monitor.
- The record from the perimeter modelling system.
- A number of spot results from surveying.

This information, supplemented by data from the plant, forms the basis for the assessment of the accident and decisions on any actions necessary.

The nature of the accident may be entirely clear. If the plant failures have been analysed in advance there may be an extremely good indication of how the accident may develop and what releases may occur. This will be an important input to the assessment because the data collected outside the plant can indicate only what has happened.

The example illustrates a number of features of data acquisition:

- The fact that the event has occurred needs to be registered as soon as possible.
- Data collection of some kind relevant to the accident should commence immediately.
- The data should be supplemented as soon as possible with data which is tailored to the event and directed towards the assessment requirements.
- Data must be collected quickly enough to be useful.

- The need for good communications must be recognized — taking account of the unusual conditions of the accident.
- Data should be recorded with all the information relevant to its interpretation and this information should be retrievable later.

(b) Assessment

The data available for the assessment have been outlined. As they become available they must be used to put together as complete a picture of the accident as possible. This will indicate how the accident may be terminated and what its impact will be on the local population. Such an assessment clearly needs the combined skills of people familiar with the plant and those knowledgeable about the effects of radiation on people.

It is not possible to assess all aspects of the accident as it occurs and specifying the aims of the assessment is essential. In our example the aim is to estimate the radiation dose to which people may be exposed. In an actual accident there could be others; most importantly, it would be vital to search for means of limiting the accident.

The actual release, measured by the discharge monitor is linked to the data from the perimeter monitor and the survey through an atmospheric dispersion model. The estimation of radiation doses from further evolution of the accident depends also upon the release model; how the release to the atmosphere is expected to continue into the future.

This atmospheric dispersion model predicts the integrated concentration in the atmosphere downwind from a unit release of material.

When a material is released to the atmosphere it moves with the wind and disperses. The rate of dispersion depends upon the amount of turbulence in the atmosphere; in very turbulent conditions dispersion and dilution occur before the material has travelled very far, but in the stable conditions prevalent on relatively calm nights the plume is narrow and therefore concentrated. The concentration at the centre of the plume depends strongly on the dispersion conditions with variations of more that an order of magnitude being possible. There are a number of sources of uncertainty: the wind direction may be wrong (admittedly unlikely) and the dispersion characteristics may be incorrectly assigned. The most frequently used model is itself known to be subject to uncertainty even if the inputs are correct (a consequence of the great complexity of the atmosphere) and it is likely to be reliable to within a factor of 2 or 3. In particular meteorological conditions and on sites with complex typography it may be in even greater error.

The perimeter monitors record the radioactive material passing over points perhaps a kilometre from the facility; members of the public will generally be some further distance away. It is, therefore, necessary to use the atmospheric dispersion model to link these two. Also, the survey results may not be available in the early stages of the accident from all points where people are living. To obtain estimates of radiation doses it may therefore be necessary to use the atmospheric dispersion model as well as the dosimetric model. This model links the amount of activity in the atmosphere to the radiation doses people receive by accounting for, for example, the movement of elements around the body and the nature of the radiation emitted.

It should be apparent that the assessment is not a matter of a few simple calculations. Typically, for accidents, there are large elements of uncertainty – not least because it may not be clear how the accident will develop.

How is the assessment made?

Managerially, a specialist team is assembled who can digest the data and deploy the models to obtain the necessary estimtes of dose.

To assist them in the use of the atmospheric dispersion model they may have packs of pre-prepared graphical information. Equally likely they will have a computer program which allows them to run the model forwards – to obtain atmospheric concentration from discharges – or backward – to obtain release estimates from measured concentrations. These computer models should remove much of the stress from assessments but it is generally unwise to regard them as more accurate. Perhaps their greatest benefit is that, with them, it is possible to look quickly at a variety of possible accident parameters – particularly important when the accident may continue with further releases under different meteorological conditions.

In coming to conclusions about the radiation doses to people the assessment team should should take account of uncertainties and should try to express them in some way in a part of their conclusions. Their assessment is to be passed to someone who has to decide on action and it could be counterproductive if, for example, too much pessimism were to be built into the analysis. Rather, best estimates should be made with explicit (if informal) statement of error bounds.

To draw a few essential points about the assessment:

- It should have closely defined aims.
- A check for consistency of data and models should be made.
- The limitations of the data and the models should be understood and be fed through the assessment procedure.
- As far as possible, detailed and complex calculations should be avoided during the emergency. Preplanning can eliminate much of this, particularly with the use of computers.

(c) Action

The results of the assessment are the basis for decisions on the actions to take to limit the effect of the accident on members of the public. It will be a statement (with some expression of confidence) of the radiation doses likely to be received by people off-site with indications of the relative importance of the various exposure pathways.

The actions which are possible are:

- Nothing.
- Advise people to shelter.
- Advise people to evacuate their homes.
- Issue stable iodate.

Sheltering is relatively straightforward and not too disruptive. It is effective in reducing radiation doses by a factor of between 2 and 10, depending upon which exposure pathways are dominant. Evacuation is, if it is done in advance of the release of activity, completely effective. If it is delayed it becomes decreasingly effective except to reduce the longer-term exposures from activity deposited on the ground. If it is done while the activity is being released it could be counterproductive. It is disruptive and may be extremely unpleasant for the sick and elderly.

An additional measure which provides protection against releases of the radioactive isotopes of iodine is the consumption of stable potassium iodate tablets. These, if taken before or during exposure, will saturate the thyroid gland with iodine and prevent further uptake of the radioactive isotopes, greatly reducing the dose. The risks associated with the consumption of potassium iodate are extremely small.

There are thus three possible courses of action (with many variations in combination and timing) and the decision on what to do will depend on the assessment. Preplanning is essential and a useful contribution to this has been the Emergency Reference Levels recommended by the National Radiological Protection Board. These indicate the levels of projected radiation dose at which each of the countermeasures should be considered. They have been arrived at by a comparison of the protection afforded by the particular measures with the degree of disruption associated with them.

Although many other factors may play a part in deciding upon appropriate protective measures these action levels provide very strong guidence. In applying them the uncertainties inherent in the dose assessment will need to be considered.

Three general observations can be made:

(a) Those responsible for taking decisions must appreciate the costs and benefits of any actions they may require.

(b) To limit the need to make judgements on complex issues during the accident, it is helpful if action levels can be worked out in advance.

(c) It must be established in advance how these decisions are to be implemented.

(d) Inform others

In the nuclear example there are a large number of organizations and people who need to be informed, in different degrees of detail and about different aspects.

Some need information because they may be involved (national government, the local community, health authorities, water authorities); others (the media, the general public) simply like to be informed. As general principles:

- Identify the groups who need to know and who would like to know.
- Tailor the information to the recipient.
- Ensure that adequate communication channels will be available.

12.7 THE EMERGENCY PLAN

The essence of effective emergency response must be to identify what can go wrong with a system and then construct a management system which can deal with the consequences of failure. Failure analysis is therefore the basis for definition of an adequate accident management capability. It can help identify the failures and their consequences but it can also indicate how it will be known when the event has occurred, how it might develop and the data which would be required for its evaluation. This can help decide to what extent the response should be automatic or tailored to deal with the specific event.

Perhaps the most important feature of emergency planning is the need to consider, in advance, the data needs and actions which might be appropriate. This is done for all nuclear installations, in great detail, as an emergency plan. This would not be appropriate for all situations but, as a general principle, the more thought put into dealing with emergencies before they happen, the more effective will be the response.

Chapter 13

DECISION SUPPORT SYSTEMS AND EMERGENCY MANAGEMENT (experiences from a case study)

Manfred Grauer
University of Dortmund
Computer Science Department
Postbox 50 05 00
D-4600 Dortmund, F.R. Germany

13.1 INTRODUCTION

The impact of normal and/or accidental release of wastes and of hazardous substances on human being and on the environment has increased enormously in recent years. This raised great attention in computer-assisted decision support for the management of waste streams and the management of emergency cases. In the paper this problem is analysed under the prospect that in these cases decision support systems (DSS) have to provide support for:

(a) dealing with multiple criteria like economic forces, technological risk, environmental considerations, etc. and

(b) the fast and robust management of complex situations.

After describing the problem (Section 13.2) in some more detail in the third section of this chapter an approach is presented to deal with multiple objectives and in Section 13.4 the DS-software system OPTIX is presented, which belongs to the so-called class of I^3-software (interactive, integrated, intelligent) and meets the second requirement (b). The use of the DSS in the case of emergency management for a plant from chemical industry is demonstrated in Section 13.5.

13.2 THE PROBLEM

Economic as well as technological considerations have dominated the evolution of industrial structures. These factors have been treated mostly separately by engineers and economists in the past. However, another major factor which has begun to have a decisive influence on the performance of industrial processes is technological risk as well as public and environmental health considerations, in particular those related to toxic and hazardous substances used in industrial production processes.

The issues of controlling process risk, waste streams, and potential environmental consequences of accidental or routine release of hazardous substances are rapidly gaining in importance as a non-negligible part in production planning and its management. To characterize the order of magnitude of those waste streams from a technological and economic point of view it is worthwhile to mention that they are estimated in [15] for the year 1986 to be 330 million metric tons. In addition to this one can find in [12, p. 58] a monetary estimation of ecological damages for one country – the Federal Republic of Germany – for the year 1986, which is given below in the following table:

Ecological damages due to:	Est. costs in Billion DM
air pollution	48.0
surface water pollution	17.6
soil destruction	5.2
noise	32.7
Σ of damages	103.5

These developments on the one hand and the availability of advanced information processing tools on the other hand have forced to take into account all these criteria (e.g. technological, economical, ecological) in the stages of the engineering design of production processes, in their control and in the corresponding emergency management problems. The paper tries to approach some of the problems under the mentioned multiple criteria and to indicate the role of I^3-decision support software for computer-aided emergency management. In Section 13.5 the application of these ideas to a production process from chemical industry under normal and exceptional circumstances (emergency situations) will be given.

13.3 THE MULTIPLE-CRITERIA APPROACH

The management problems as stated earlier are multiple-criteria ones. To deal with this multiplicity of the criteria an approach is needed, which transforms the originally multiple-criteria problem into a scalar-valued one to analyse it on a computer. Such a transformation (function) has to have the following features to serve for decision situations under analysis:

- it should be similar to utility functions and its maximization should lead to efficient decisions relative to the actual list of criteria;
- it should depend on the aspirations of the decision maker and thus expressing the current preference structure among the criteria;
- it should correspond to the minimization of a distance of decision outcomes to aspiration levels, if the latter are not attainable, and the maximization of such a distance, if they are attainable;
- the test of attainability and efficiency of aspirations should be numerically easy; and
- it should be also usable in the case of dynamic outcomes, which is in the emergency situation most important.

For such a transformation (function) the idea of satisficing behaviour and its mathematical formalization for decision support is used as developed in [13, 14]. To give a brief understanding of this approach for the purposes of the current analysis it is assumed that, the decision space $E_u = R^n$ and the set of admissible decisions

$U_0 \in R^n$ are compact. Let there be p criteria or objectives of interest to the decision maker, $E_q = R^p$, and let the outcome mapping $f : U_0 \rightarrow R^p$ be continuous, hence the set of attainable outcomes $Q_0 = f(U_0)$ are also compact. Under the assumption that the decision maker wants to maximize all outcomes, the partial ordering of the outcome space is implied by the positive cone $D = R_+{}^p$. In [13] is stated if the cone D is closed and the set Q_0 is compact, then there exist D-efficient or D-optimal elements of Q_0. These are such elements $q' \in Q_0$ that

$$Q_0 \cap (q' + D') = \emptyset$$

where D′ is the cone D except its origin and q′ + D′ denotes the set D′ shifted by q′. The corresponding decisions $u \in U_0$ such that $q' = f(u')$ are D-efficient as well. The set of all D-efficient outcomes is then:

$$Q'_0 = \{ q' \in Q_0 : Q_0 \ (q' + D') = \emptyset \} \ .$$

For applications in decision support systems a substitute scalarizing function is introduced, the so-called achievement function, which is the above mentioned transformation (function) from the p-dimensional objective space into the one-dimensional space: $s : R^p \rightarrow R^1$. This function is guided by reference levels or aspirations of the decision maker about target values or trajectories for the criteria $(\bar{q}_i)$ and its maximization leads to D-efficient decisions. In the multiple-criteria dynamic case as needed and discussed here for emergency management this function can be written as:

$$s(q, \bar{q}) = \min_{t_i \leq t \leq t_k} \ \min_{1 \leq i \leq p} \ \mu_i [q_i (t) - \bar{q}_i (t)] + \dots$$

$$\dots + \delta \sum_{t=t_0}^{t_f} \ \sum_{i=1}^{p} \mu_i [q_i (t) - \bar{q}_i (t)], \tag{1}$$

where $\delta > 1$, $\mu_i \geq 0$ and $\sum_{i=1}^{p} \mu_i = 1$.

The solution of the multiple-criteria emergency management problem consists now in finding those decisions which maximize (1) for given reference levels $\bar{q}_i$ under the constraints represented by the dynamical mathematical model of the industrial process and corresponding lower and upper bounds on control-, state-, and output-variables. This is in principle a general nonlinear programming problem with an iterative procedure (normally the solution of a system of differential equations) in the constraints and in the objective function. To solve this problem in a reliable manner and sufficiently fast in most cases (excluding academic examples) a combination of random and direct search methods with derivative methods would be needed (see [11]). To make these algorithms available from a library of standard programming methods in a decision support system is the subject of the next section.

13.4 OVERVIEW TO THE I^3 - DECISION SUPPORT SOFTWARE

The system as described now exists in the Computer Science Department of the University of Dortmund as a student-version called ISAAC for the use on ATARI-computers and serves mainly for educational purposes and as a professional version called OPTIX for the use on SUN-computers for large-scale problems. In [10] the questions of designing a corresponding user interface for the DSS are investigated in some more detail. The main idea there was to use the "WYSIWYG" (what you see is what you get)-concept from desktop-publishing in combination with an object-oriented computer language. The Smalltalk System [7, 8] is such a combination of an object-oriented computer language with a graphic user interface and various programming tools such as debugger and a system-browser. When developing ISAAC and OPTIX that type of user interface was taken as the model for these DSS. This concept together with the above mentioned multiple-criteria approach stands for the first " I^1 "(interactive).

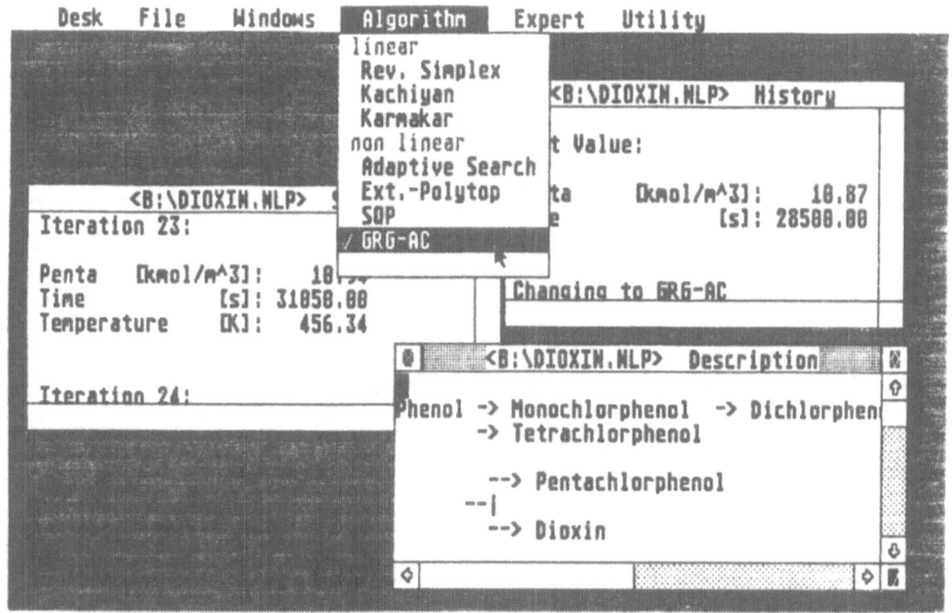

In addition to that in the DSS the idea of integration of several different linear and nonlinear programming methods stands for the second " I^2 " (integrated). On the other hand the package was meant to be an attempt to combine symbolic and numeric methods , e.g. using a knowledge-based system for the choice of the best available programming method and/or computing symbolic derivatives ("I^3" intelligent).

Both DSS (ISAAC, OPTIX) consist of a window interface, a language for problem definition and some optimization algorithms like Sequential Quadratic Programming (SQP) or Generalized Reduced Gradients with Active Constraints (GRG-AC).

The above diagram, which is a hardcopy of the screen from an ATARI-computer shows all the currently implemented algorithms together with a problem description and a test run with some iteration values.

The use of more than one algorithm in the nonlinear case is based on the fact that there is no one single method which can solve all optimization problems like (1) in the best possible way. The algorithms must therefore be chosen in an intelligent fashion, either by the user and/or with the support of the knowledge-base.

After years of experience with developing DSS in the language FORTRAN-77 these systems were developed from scratch in the computer language C, for the reason that C is probably the fastest high level language necessary for a comfortable man-machine interaction. Furthermore some of the software tools like LEX, YACC and SunTools which are needed can only be used in combination with respectively C or the operating system UNIX. The use of C also allows high portability especially under the UNIX system.

It was decided to develop a context-free language for the problem description because this can be done very quickly using as a generator for a scanner LEX and as a generator for a parser YACC (see [10] for details). The problem description language is descriptive rather than imperative so it is neither possible nor necessary to program the problem solving method.

To deal with the multiple criteria case the satisficing approach is implemented and the user guides the problem solution by setting his aspiration levels as described above. To get a fast and robust problem solution as needed in emergency cases the system supports the parallel and sequential use of different algorithms for one problem. If there is a multiprocessor-system at hand a physically distributed problem solution can be realized. The main idea for that approach is to use a knowledge base with different patterns of the problem structure (like multimodality far from solution and a quadratic model in the vincinity of the solution) and by that to control from a higher hierarchy level the serial and/or parallel work of several algorithms. In optimization (the solution of problem (1)) this means for instance that one can deal with the problem of getting the global solution, with nondifferentiabilities or with a quadratic approximation in the neighbourhood of a solution.

What was presented here in this section is a general DSS-module, which meets several requirements to support emergency management. For a specific application problem-related modifications have to be made. How this can be done, will be demonstrated in the next section.

13.5 A CASE STUDY FROM THE CHEMICAL INDUSTRY

The chemical process of direct chlorination of phenol raised attention in connection with the "Seveso"-accident in Italy in 1976. A simplified technological schema of that process is given in Figure 13.1 [4, 5, 9, 11].

The extremly toxic 2, 3, 7, 8 - tetrachlorodibenzo-p-dioxin and 2, 3, 4, 7 - pentachlorodibenzofuran are byproducts of that process. The process representation as well as the process risk description is given in more detail in [9, 11]. Thereby it is assumed that the chlorination of phenol(0) proceeds stepwise so that five chlorophenols of commercial value are obtained: monochlorophenol (1), dichlorophenol (2), trichlorophenol (3), tetrachlorophenol (4) and pentachlorophenol (5).

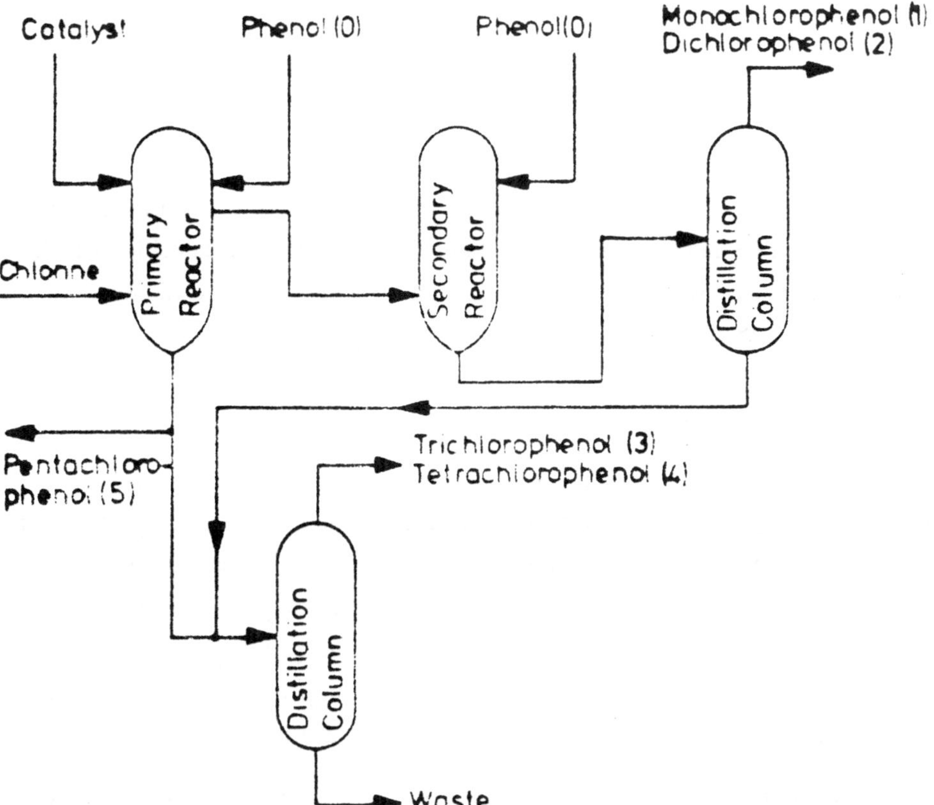

Figure 13.1. Simplified technological schema of the chemical plant for the direct chlorination of phenol [9].

The subprocesses and substances involved in that manufacturing are summarized below:

phenol(0) + chlorine → monochlorophenol(1) + HCL

monochlorophenol(1) + chlorine → 2,4-dichlorophenol(2) + HCL

2,4-dichlorophenol(2) + chlorine + (ALCL3) →

→ 2,4,6-trichlorophenol(3) + HCL

2,4,6-trichlorophenol(3) + chlorine + (ALCL3) →

→ 2,3,4,6-tetrachlorophenol(4) + HCL

2,3,4,6-tetrachlorophenol(4) + chlorine + (ALCL3) →

→ pentachlorophenol(5) + HCL

pentachlorophenol(5) + chlorine + (ALCL3) → dioxin(6).

The numbers in parentheses are used as abbreviations for the substances in the paper, and ALCL3 indicates a catalyst (anhydrous aluminium chloride) which is added to the reactor when the chlorination has proceeded to the dichlorophenol stage (see Figure 13.1).

The problem of control and management of this chemical plant in a normal situation consists of meeting the demand for the five products (pentachlorophenol is the main product, but mono-, di-, tri- and tetrachlorophenol are also products) and/or maximizing the target products in a time-minimal production period. Here the interdependencies between the different outputs of the process have to be taken into account. That means for instance that the amount of pentachlorophenol is a result of all steps of the reaction and any intermediate product would reduce its throughput. This requests to understand the production profile as a compromise of all target products which meets best the current demand profile.

Moreover to these production criteria the amount of toxic substances (e.g. dioxin) has to be minimized or at least kept below certain upper bounds. This leads in the normal production mode as well as in non-normal cases (e.g. runaway reaction, operator error, equipment failure, fire, explosion etc.) to the necessity to deal with a dynamic multiple-criteria decision problem.

On the other hand these mathematical models have already been used in the design stage when the decision about the principal process chemistry is made and the engineering design of the layout of the plant is done. Of course the decision variables are then different from those in the case of production planning for a given plant. But as is well known, in the design stage about 80% of the economic efficiency is determined and also the possibilities to meet ecological and emergency requirements. This calls for an integration of decision support for computer-aided management in all these stages.

For the chlorination process the theoretically and experimentally verified conceptual model is that of a consecutive reaction from phenol(0) over pentachlorophenol(5) to dioxin(6) of the following type:

$$(0) \rightarrow (1) \rightarrow (2) \rightarrow (3) \rightarrow (4) \rightarrow (5) \rightarrow (6)\,.$$

The mathematical formulation of this concept is a system of ten differential equations(DE) consisting of:

7 DE for the mass-balances of all substances 0 through 6,

1 DE for the mass-balance of chlorine in the liquid phase,

1 DE for the mass-balance of chlorine-consumption and

1 DE for the energy-balance in the reactor.

In addition to the system of differential equations, there is also a set of highly nonlinear algebraic equations relating the interdependencies of densities, reaction coefficients, heat transfer coefficients, etc. with time, temperature, pressure and material composition, e.g.: concentration.

The mathematical description of the process is presented here in such detail to give an understanding of the process, and finally to solve the control/optimization task as part of a decision problem in an emergency case.

The mathematical model to describe the dynamic behaviour of such continuous or semicontinuous (batch-wise) production processes as given above is usually a system of differential equations as follows:

$$\frac{dx(t)}{dt} = F[x(t), u(t), t], \quad x(t) \in R^m, \quad u(t) \in R^n \tag{2}$$

with the given initial state of the production system at the time $t = t_o$:

$$x(t_o) = x_o \qquad t \in [t_o, t_f] \tag{3}$$

where x is the state and u the control variable and $[t_o, t_f]$ indicate the time horizon.

Usually the state- and the control-variables are subject to further constraints:

$$g[x(t), u(t), t] \leq 0 \quad , \tag{4}$$

and

$$h[x(t), u(t), t] = 0 \tag{5}$$

It is assumed that the performance measure has the form of the following scalar-valued functional:

$$\min \left\{ \int_{t_o}^{t_f} f[x(t), u(t), t]dt + \Phi[x(t_f)] \right\} \tag{6}$$

The functional (6) represents the time-dependence of the criteria of the decision processes like economic ones or considerations in the emergency case. Under normal production conditions this can be an overall cost-function. For decision problems connected with abnormal situations the integral can represent the amount of toxic substances for which the danger exists to be accidentally released and which has to be minimized.

The most common approach to solve this class of problems is to convert such an optimal control problem (2)—(6) into a static nonlinear programming problem. This can be done by approximating the continuous time-function u(t), i.e. the control-trajectory (or policy) by a discrete time-function. The new decision trajectory or decision policy is derived by choosing an appropriate structure with finitely many free parameters and to consider the state variables x(t) as functions of the control variables through the differential equations (2). Hereby the problem is solved in the reduced space of controls, and the state variables are obtained by the solution of an initial value problem.

Following this approach the interval $[t_o, t_f]$ is partitioned into a (not necessarily equidistant) time-grid:

$$t_o = t_1 < t_2 < \ldots < t_j < \ldots < t_k = t_f, \tag{7}$$

and on this grid the finite set of control variables is defined:

$$u(t) = [u_1(t_1), \ldots, u_n(t_1), u_1(t_2), \ldots, u_n(t_{k-1})]. \tag{8}$$

The infinite control or decision problem (2) – (6) can then restated in the following way:

$$\frac{dx(t)}{dt} = F[x(t), u_i(t_j), t], i = 1, \ldots, n; j = 1, \ldots, k-1 \tag{9}$$

with $\quad x(t_o) = x_o \qquad t \in [t_o, t_f]$ (10)

$$g[x(t), u_i(t_j), t] \leq 0, \; i = 1, \ldots, n; \; j = 1, \ldots, k-1 \tag{11}$$

$$h[x(t), u_i(t_j), t] = 0, \; i = 1, \ldots, n; \; j = 1, \ldots, k-1 \tag{12}$$

and

$$\min \{ \int_{t_o}^{t_f} f[x(t), u(t), t]dt + \Phi[x(t_f)]] \tag{13}$$

The nonlinear single-criterion programming problem then is to find the piecewise-constant control strategy (8) which minimizes (13) under the constraints (9) – (12). This is a nonlinear programming problem with an iterative procedure in the constraints and in the objective function.

The current prototype module of the DS-software-system supports a single user in the case of a non-normal situation of the plant when the operator has to meet the production goals and/or to satisfy the safety requirements in an emergency case.

Figure 13.2 shows a simulation run of the chemical plant for one specified efficient decision case. One such simulation run for a result as specified in Figure 13.2 takes on an IBM-PC XT/AT with an APU-8087 about 40 to 50 minutes. This represents about 20 to 30 function evaluations for the underlying nonlinear programming problem. The same problem is solved on a VAX 11/780 in less than 10 minutes and on a SUN 3/260 which is the target machine in less than 30 seconds.

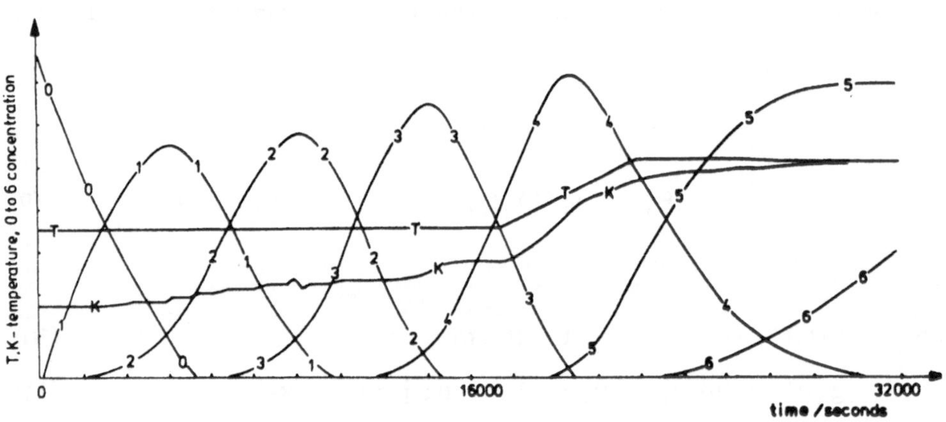

Figure 13.2. Process dynamics of a chemical plant for the direct chlorination of phenol (T, K - temperature in the reactor and of the coolant in °C and (0), (1), (2), (3), (4), (5), (6)- concentration in kmol/m^3 of phenol, mono-, di-, tri-, tetra-, pentachlorophenol, dioxin) after optimization (efficient decision case).

13.6 CONCLUSIONS

The concept of this prototype decision support software system is that it is not only applicable to the specific process under discussion here, but to a broader range of integrated production planning and the management of emergency cases.

The use of decision support software in emergency situations is already known for high-order corporate crisis management (e.g. [1]) and in industry for the management of nuclear power plants (e.g. [3]). For the application as a decision aid in on-line production management the step has to be made from the presented off-line type system with a so-called "deep" model to "shallow" mathematical models. This means the integration of knowledge processing not only into the solution process but also for the updating of the mathematical description of a plant by on-line information from process measurements into DSS [2]. The development of information and communication technology and sensor technology supports this step already and research in this direction is underway (e.g. [6]).

ACKNOWLEDGEMENT

I would like to thank my students from the Computer Science Department of the University of Dortmund, who supported the development of the software-systems ISAAC and OPTIX.

REFERENCES

1 Andriole, St., "Microcomputer-based decision support for high-order corporate crisis management", in "Microcomputer decision support systems: design, implementation and evaluation", St. Andriole (Ed), North Holland, 1986.

2 D'Ambrosio, B., Fehling, M.R., Forrest, St., Raulefs, P. and Wilber, B.M., "Real-time process management for materials composition in chemical manufacturing", IEEE Expert, pp. 80-93, Summer 1987.

3 Embry, D. and Humphreys, P., "Support for decision making and problem solving in abnormal conditions in nuclear power plants", in "Knowledge representation for decision support systems", L.B. Methlie and R.H. Sprague (Eds), Elsevier, North Holland, 1985.

4 Fedra, K., "Advanced decision-oriented software for the management of hazardous substances, part I: structure and design", CP-85-18, International Institute for Applied Systems Analysis, Laxenburg/Austria, 1985.

5 Fedra, K., "Advanced decision-oriented software for the management of hazardous substances, part II: a demonstration prototype system", CP-86-10, International Institute for Applied Systems Analysis, Laxenburg/Austria, 1986.

6 Gallanti, M. and Guida, G., "Intelligent decision aids for process environments: an expert system approach", in "Intelligent decision support in process environments", E. Hollnagel et al. (Eds), NATO ASI Series, Vol. F21, Springer-Verlag, Berlin, Heidelberg, 1986.

7 Ganzinger, H., Heeg, G., Baumeister, H. and Ruger, M., "Smalltalk-80, in Informationstechnik" – it, 29, Heft 4, Seite 241-251, 1987.

8 Goldberg, A., "Smalltalk-80: The interactive programming environment", Addison-Wesley Publishing Company, Reading, 1984.

9 Grauer, M. and Fedra, K., "Intelligent decision support for technology assessment: the chemical process industry", in "Toward interactive and intelligent decision support systems (Vol. 2)", Y. Sawaragi, K. Inoue and H. Nakayama (Eds), Springer-Verlag, 1987.

10 Grauer, M., Albers, St. and Frommberger, M., "Concept of a window-based interface for SIAD", Proceedings of the 27-th Meeting of the European Working Group on MCDA, Mons, March 1988.

11 Grauer, M., "A case study in the use of mathematical models for decision support in production planning", in "Mathematical models for decision support", G. Mitra et al., NATO-ASI Series, Springer-Verlag (forthcoming), 1988.

12 Seidel, E. and Menn, H., "Die ökologische Lage und die Haltung der Betriebswirtschaft bei ihrer Bewältigung", internal report of the University of Siegen, Siegen, FRG No. 4, 1987.

13 Wierzbicki, A., "A mathematical basis for satisficing decision making", Mathematical Modelling, No. 3, pp. 391-405, 1982.

14 Wierzbicki, A., "On the completeness and constructiveness of parametric characterizations to vector optimization problems", OR Spektrum, 8:73-87, 1986.

15 "World Resources 1987", a report by The International Institute for Environment and Development and The World Resources Institute, Basic Books Inc., New York, 1987.

Chapter 14

SAFETY INTEGRITY MANAGEMENT USING EXPERT SYSTEMS

Dr Peter Andow
KBC Process Automation
Chilworth Research Centre
Southampton U.K.

14.1 INTRODUCTION

Modern process plants are large and complex. When failures occur the consequences can be extremely serious in terms of both economic loss and human mortality.

The public at large are justifiably concerned about the hazards imposed upon them by the chemical and nuclear industries. In the Western world public concern about safety issues has caused a response from industry in the form of much stronger internal awareness of safety issues. There has also been a response from government in the form of more regulatory legislation. (The most significant manifestation of public concern is the current state of the nuclear power industry in the USA – where the Three Mile Island incident has had an enormous impact).

14.2 SAFETY AND RISK ANALYSIS

This increase in awareness of process hazards has led to a wide acceptance of the need for improved safety at the **design** stage. In the UK (but not in every country), the Hazard and Operability Study (HAZOP) [2] is widely used to:

- Identify hazards
- Modify the design to:

  (a) eliminate the hazard; or

  (b) reduce its rate of occurrence; or

  (c) reduce its consequences.

HAZOP will usually reveal a range of potential hazards. Most are dealt with by design modifications as noted above. These methods rely on standard "good practices" and design codes that are well-known.

A much smaller number of hazards are then analysed in more detail using probabilistic methods to estimate the rate of occurrence and the magnitude of the associated consequential effects. Methods used for this (in increasing order of popularity) are:

(a) Fault Trees

(b) Event Trees

(c) Cause Consequence Diagrams and other methods.

14.3 THE EFFECTS OF APPLYING SAFETY AND RISK ANALYSIS

The use of the methods mentioned above give the potential for major improvements in safety. Safety improvements are only realized if plants are managed properly:

- The general management must be competent and also be genuinely committed to safe operation
- Protective systems must be maintained and tested according to the assumptions made in the analyses

- Standards must be maintained:

  (a) the plant should be tidy

  (b) protective equipment kept in good condition

  (c) documentation kept up-to-date

  (d) training needs identified

  etc.

Even when all of these conditions are satisfied some problems remain. In particular, the protective systems may not be very well understood by the operating team.

Consider the accident at Three Mile Island (TMI) – the protective systems were automatically (and correctly) initiated. The operators did not understand why water was being pumped into the reactor and so they stopped the pumps.

A simplified profile of the incident shows its essential form – see Figure 14.1.

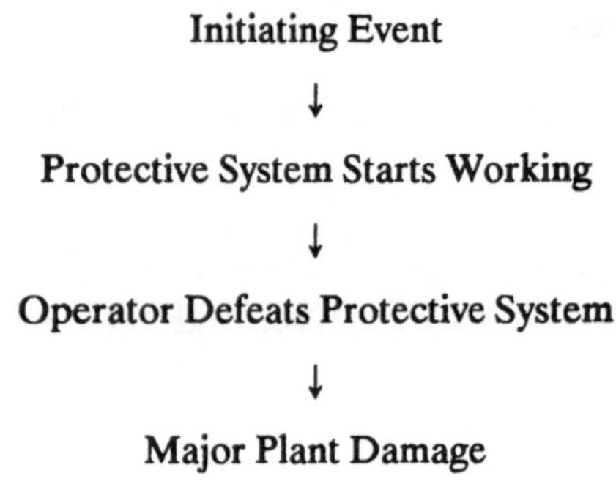

Figure 14.1. A profile of the TMI accident.

In this extreme case the accident was made far worse by human intervention.

It is quite clear that there is no point in providing a complex and expensive protective system if it is likely to be defeated by the operator. In this chapter we examine the role of a knowledge-based system which overcomes this problem –

and other related problems. See [1] for a more general discussion of knowledge-based systems in safety.

14.4 SAFETY INTEGRITY MANAGEMENT

In a case like that at TMI the plant safety can be improved by making the protective systems more "transparent". The reason for pump initiation could be given to the operator. The explanation must be presented clearly. The operator must be able to verify the information given and be able to delve deeper for more basic causes.

An aid designed for this role would clearly need structural knowledge of the protective systems. This knowledge would be encapsulated at design time.

This type of aid would be useful for a small number of incidents – ones with a profile like that at TMI. We must note that it would not help at all in most others (e.g. Flixborough and Bhopal).

Consider a profile of the Bhopal disaster as shown in Figure 14.2.

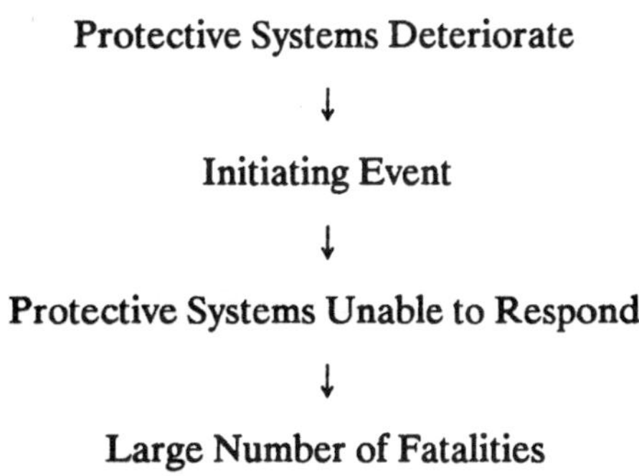

Figure 14.2. A profile of the Bhopal disaster.

This is again an extreme case – both in terms of the number of fatalities and in the scale of "deterioration" of the protective systems. The Bhopal incident shows that it is futile relying on safe design if operational practices and maintenance are not effective.

One feature of Bhopal that is a contributory factor in many incidents is that there is some deterioration in protective systems – but rarely on such a scale as at Bhopal.

It must be stressed that some deterioration is inevitable – even in well-managed plants. this is because protective systems are, by their nature, passive for most of the time. The designer recognizes this and specifies the required testing and test intervals as a means of discovering faults that would otherwise remain unrevealed.

A knowledge-based system can be used as an aid to highlight the importance of failures in passive systems. Protective systems may be thought of as "barriers" that stop initiating events from causing serious failures – see Figure 14.3. The figure shows two barriers. Sometimes there is only one and sometimes more than two, but two is probably most common.

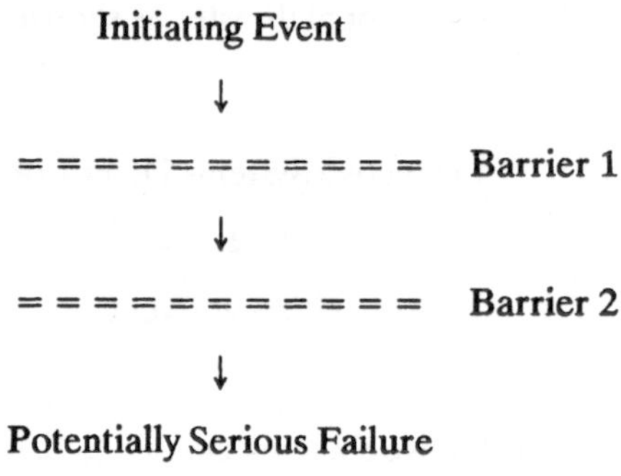

Figure 14.3. The "barrier" model of protective systems.

For high-hazard systems there are typically a whole series of initiating events with potentially serious consequences. There are likely to be different protective devices that are appropriate for the various events. For many events the human operator is the final barrier – or "last line of defence". The operator is often included because he is seen as being flexible and may be able to cope with conditions not foreseen by the designer – or slightly different from the conditions predicted by the designer.

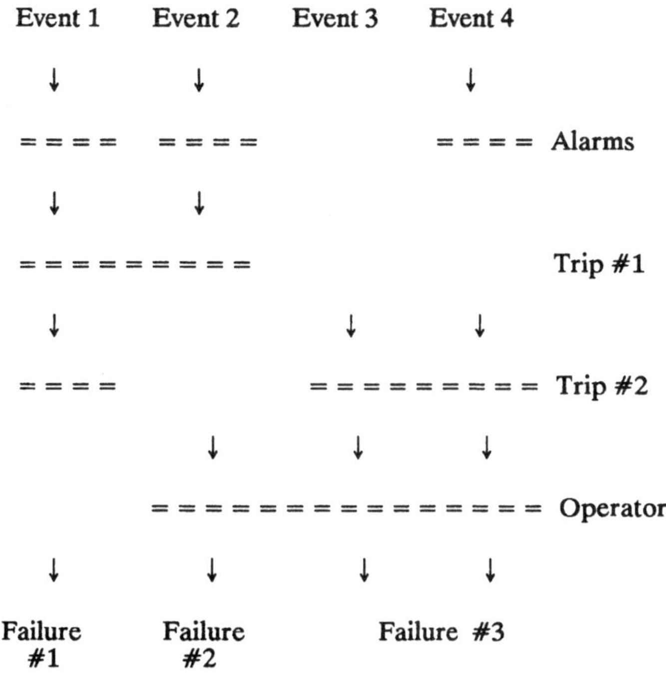

Figure 14.4. Diverse events and protective systems.

Figure 14.4 shows how a set of different initiating events are dealt with by a set of protective systems.

On most it is clear from Figure 14.4 that a failure of Trip #1 is particularly significant for Event 2 whilst failure of Trip #2 is most significant for Event 3. Ideally, when a failure is discovered (by testing) then the impact of this on system integrity would be recognized by the operating team. In practice – particularly on a complex plant with difficulty in recognizing which initiating events are most affected by the failure.

A knowledge-based system can be used to continuously monitor the safety integrity. The knowledge-base contains details of the various initiating events, their potential consequences and the protective systems that are appropriate. The system could also contain details of the test schedule and could then prompt the maintenance supervisor for the results of the scheduled tests.

14.5 KNOWLEDGE -BASE CONTENTS

The knowledge-base would contain the following:

- The list of initiating events, their potential consequences and the appropriate protective systems.
- Inputs from the process monitoring system so that protective system activation can be monitored and analysed (for explanation to the operator).
- (Optionally) Details of protective system test schedules and the date of the last successful test.

14.6 SUMMARY OF SYSTEM FUNCTIONS

The system would be able to :

- Monitor protective system.
- Draw the operator's attention to Initiating Events that are dramatically affected by particular failures of the protective systems – as these are discovered by testing. This effectively updates the safety analyses in the light of known failures.
- Monitor the testing of protective systems and even update reliability calculations as systems are "proved" by testing.

14.7 DISCUSSION

The knowledge-based system proposed here is not claimed to be suited to improving safety in all circumstances. The ability to explain protective system activation is appropriate to failures such as that at TMI. The ability to monitor and warn the operator of safety system degradation as successive failures occur is appropriate to many plants – because of the requirement to demonstrate (by testing) whether normally passive systems are still capable of working. It would be dangerous to assume that such a system might have prevented the disaster at Bhopal – where the large number of failures must have been known by operators and management.

One response to some of the mistakes made by operators is to suggest that control systems should be completely automated. The author believes that this would be counter-productive. Good operators are a valuable asset on many plants and have often performed well in emergency conditions. The system proposed is intended to enhance the reliability of operator responses as they are faced with ever more complex systems.

ACKNOWLEDGEMENTS

The author wishes to acknowledge permission from the directors of KBC Process Automation for permission to publish this article. All correspondence should be addressed to Peter Andow, KBC Process Automation, Chilworth Research Centre, Southampton, SO1 7NP, UK.

REFERENCES

1 Andow, P.K. and Ferguson, G., "Expert Systems and Chemical Process Safety", Proc. World Bank/AIChE/EPA Conference on "Preventing Major Chemical Accidents", Washington D.C., USA, 1987.

2 Kletz, T.A., "HAZOP and HAZAN" (2nd edition), available from the Institution of Chemical Engineers, 165-171 Railway Terrace, Rugby, CV21 3HQ, UK.

Chapter 15

POWER SYSTEM ALARM ANALYSIS AND FAULT DIAGNOSIS USING EXPERT SYSTEMS

P.H. Ashmole
Central Electricity Generating Board

15.1 INTRODUCTION

One major application of expert systems is in the field of Alarm Analysis Fault Diagnosis. The problem arises wherever a complex control operation of a large process is involved. Power system control is one such problem area.

The Power system control problem is characterized by the very large number of potential alarms which could arrive at a Control Centre. In the UK each control centre controls typically 80 substations (connecting points on the 400/275 kV supergrid system) and each substation has the potential to initiate many hundred different alarms. Major fault conditions initiate only a small proportion of these alarms and occur only rarely – say less than once per year in a particular control room, but when they do occur they present a number of problems to the operator:

- He is in danger of being swamped by alarm data to such an extent that he has difficulty in sorting out the important alarms and the basic cause of the fault.
- By definition a major disturbance will be an unplanned disturbance since power systems are normally operationally planned to operate such that they can withstand a fault on any single plant item (and possibly any two plant items) without a major system disturbance involving loss of consumer supplies. Such major disturbances normally occur following an unusual and unexpected set of events which may be difficult for the operator to understand and diagnose.

- The operator's main objective in handling a major disturbance is to understand the cause of the fault and appropriate remedial action as soon as possible. Present control and alarm presentation systems are not well geared to this objective.

For many years the CEGB have sought to develop techniques to rationalize the alarms presented to the system control engineer since this allows direct operation of substation switching from the Area Control Centre rather than through verbal commands to a substation operator, but a satisfactory solution to the problem has only emerged with the possibility of using an expert system approach.

The term expert system covers a large range of techniques and normally implies a knowledge base which is acquired by questioning the expert. In the case of power system alarm analysis the problem is rather different since a high proportion of the knowledge can be logically derived from an examination of the process involved. The result is a particular structure for the proposed system based on an hypothesis selection followed by a simulation to verify the hypothesis. This type of structure is common to all power system alarm analysis expert systems which have so far been developed and are discussed below.

15.2 EXPERT SYSTEMS FOR POWER SYSTEM ALARM ANALYSIS ALREADY DEVELOPED

Table 15.1 indicates some expert systems for power system alarm analysis that have already been developed and for which preliminary information has already been published (references 1, 2, 3, 4). None of the systems are in full commercial use and only the EVS system is currently available to the power system operator on a trial basis. An examination of these systems has enabled us to appreciate the problems in such a development and formulate our own specification. The main problem areas with these systems which prevent user acceptability are:

- Poor success rate in recognizing and diagnosing fault situations.
- Poor user interface.

- Inability to deal with multiple fault situations.
- Presentation to the operator of spurious fault information rising from routine or maintenance system switching activity.

15.3 EXISTING SUBSTATION CONTROL ARRANGEMENTS

At present each area grid control centre (ACC) controls some 80 substations.

Enquiries into the "cause" of a fault are primarily satisfied by alarm data. Information flow and content from alarm source to the ACC differs according to the method of substation control. Three categories of control exist:

(a) Locally Operated Sites

Automatic data transmission to the ACC comprises: switch status, auto reclose operated and reset and overload alarms.

Alarm data is reported to the ACC verbally, by telephone, either on request or in accordance with agreed procedures. The information basically comprises substation annunciator legends but this can be supported with additional facts such as relay flag indications, local weather conditions, visual condition of the plant.

(b) Remote Operated Sites

Remote operated sites are unmanned and controlled from an adjacent manned substation. Automatic data transmission to the ACC and verbal reporting of annunciator legends is similar to locally operated sites. Further detailed alarm information is not available without a site visit and hence further delays in communication occur.

(c) Operation direct from ACC

Automatic data transmission is as above plus plant alarms. Alarm data, typically over 100 items, are transmitted to the ACC. The substation is not equipped to rationalize and hence much spurious data is allowed through, e.g. fleeting alarms.

Studies of recent fault situations indicate that around 60% of alarms transmitted from such sites are not required for system decision making, and further implementation of direct operation from the ACC awaits the introduction of expert system alarm analysis.

15.4 DISCUSSION OF ALARM DATA FLOW

Different amounts of data are received at the ACC according to the method of substation control. Metering errors are mainly associated with analogue values.

During a recent system incident, a peak alarm activity of 85 alarms per minute was received at the ACC.

Operation direct from the ACC, which will be introduced widely when the expert system is available, increases the amount of data received approximately fourfold. The Control Engineer has to search a vast amount of alarms to determine the "cause" and "effect" of the system incident.

15.5 EXPERT SYSTEM REQUIREMENTS

The overall objective of the expert system is to provide the control engineer with higher quality, reduced quantity information; enabling him to understand more quickly the effect of the fault on the system, to assess the state of the system and to determine remedial actions. Consequently, the output of the expert system needs to be clear and concise, and its operation should demonstrate reliability and robustness.

15.6 USER INTERFACE

It is difficult to prescribe a fixed user interface for the system, because the information provided should ideally vary with the number of faults or the state of the system:

- The first level of data presentation should provide a single message on an incident list to replace associated switchgear status changes and alarm messages.
- The second level of presentation should be a list of status and alarm information used.
- The third level should be the rules used to recognize the fault (i.e. the system should be able to explain how it reached its conclusion).
- The display should be integrated into control room facilities.

Table 15.2 is an example of the alarms received for a recent fault (3rd April 1987). The log shows the quantity of data received and the time taken to receive it; the actual time taken for the fault and auto-reclosure was probably about 30 seconds. These messages could be replaced by a simple statement such as:

"Transient fault on AXMI-MANN, successfully auto-reclosed."

The prime message of interest to the Control Engineer will be this first level message. The second and third levels of information would not necessarily be used by the Control Engineer but would provide the ability for more detailed inspection of information for testing, development, post-fault analysis and would be available to the Control Engineer if required.

15.7 REQUIREMENTS UNDER DIFFERENT FAULT CONDITIONS

The system provided must be able to recognize, handle and diagnose multiple faults.

Faults generally fall into three categories; single faults, multiple faults and system incidents and all categories must be handled by the expert system technique. For all faults, the output of the expert system should present the cause of the fault i.e. a rationalization of the status changes and alarms that are received, as a summary message together with any non-standard alarm behaviour and unresolved information.

15.8 DATA STRUCTURE

In addition to the stringent user requirements discussed above further problems related to the data structure makes Power System Alarm Analysis particularly difficult:

- The data is obtained from outstations by scanning the alarm initiating devices. Data relating to a particular fault condition can arise from the scanning of several substations and the order of data arrival at the ACC is not necessarily the same as the order of initiation.
- The data is not time tagged so timing information cannot be used in the analysis.
- Some fault sequences can last for a lengthy period of say several minutes particularly where delayed auto-reclose is involved, so the alarm data analysed at a given time may be incomplete.
- During a particular fault it is common to get incorrect equipment operation, e.g. switches fail to operate or operate unnecessarily following a fault condition.

Any alarm analysis system has to deal with the above issues and still provide a very high standard of performance.

15.9 EXPERT SYSTEM STRUCTURE

The expert system appropriate to the power system alarm analysis problem is expected to have the following components:

- A hypothesis generator which formulates a series of fault hypotheses in order of likelihood. This hypothesis generator would be based on a number of relatively simple rules.
- A System Simulator which simulates the fault conditions suggested by the hypothesis and indicates the resulting alarm conditions.
- A comparator which compares the alarm conditions generated by the simulator with the actual alarm condition and indicates the correctness or otherwise of the hypothesis.

15.10 DEMONSTRATOR PHASE OBJECTIVE

The first phase of the project will be to construct a demonstrator. This is expected to take some two years and have the following objective:

- Establish methods: knowledge acquisition and knowledge representation.
- Analyse and understand problem domain.
- Address particular technical issues e.g. uncertain information, probabilistic inference.
- Establish system architecture.
- Propose interface methods (and experimentation).
- Establish methods for refinement and validation.
- Identify full-scale issues.
- Identify tool support for subsequent phases.
- Demonstrate cost-effectiveness.
- Demonstrate key functionality/capabilities.

Table 15.1

POWER SYSTEM ALARM ANALYSIS

AND FAULT DIAGNOSIS

Expert Systems

Some Developments outside CEGB

ORGANIZATION	APPLICATION	STATUS	COMMENTS
EVS Stuttgart	Alarm Analysis of 380kV Network	In Trial operation in Control Centre	Simple Pattern Recognition system Poor successful analysis less than 90%
Kyushu Japan	Alarm Analysis of 220/100kV network	Demonstration system adjacent to Control Centre	Gives fault probabilities to operator. Little known about opeator acceptability.
EdF	Alarm Analysis of EdF 380Kv network.	Demonstration phase. Commissioned 1992.	Still in development phase. Major resource commitment aiming at 100% successaful analysis.
Carnegie Melon/ Alleghenny Power System	Alarm Analysis	University Research Phase	Conceptual design interesting
Thames Polytechnic/ Southern Board	Alarm Analysis of Distribution Network	Demonstration in Polytechnic (Alvey funded)	Not likely to be a adopted by Southern Board. Insufficiently developed for major system application.

Table 15.2

ALARM LOG FOR MANNINGTON FAULT

3/4/87

2013	ABTH	GT7	OPEN
2022	1 MANN4	AXMI MAIN PROT	OPERATED
2022	MANN4	X220 /	OPEN
2022	AXMI4	X120 /	OPEN
2022	1 MANN4	NON-URGENT ALARM	ON
2022	1 CHIC4	MANN CCT MAIN PROT	OPERATED
2022	1 CHIC4	MANN CCT PHASE U/B	ON
2022	1 CHIC4	EXET CCT PHASE U/B	ON
2022	1 MANN4	AXMI I/TRIP RECD	ON
2022	AXMI-SGT	2 132 VOLTS	LOLIM VIOLATED 134 ON
2022	AXMI1	180 /	OPEN
2022	1 CHIC1	SGT2 CCT PHASE U/3	ON
2022	1 MANN4	DIESEL FLT	ON
2022	1 CHIC1	SGT1 CCT PHASE U/3	ON
2022	1 MANN4	DIESEL FLT	OFF
2022	1 CHIC4	MANN CCT MAIN PROT	RESET
2022	1 MANN4	FAWL PHASPE UNBALANCE	ON
2023	1 CHIC1	SGT1 CCT PHASE U/3	OFF
2023	1 MANN4	LOVE PHASE UNBALANCE	ON
2023	1 CHIC1	SGT2 CCT PHASE U/B	OFF
2023	1 MANN4	AXMI PHASE UNBALANCE	ON
2023	1 CHIC4	MANN CCT PHASE U/B	OFF
2023	1 MANN4	AXMI MAIN PROT	RESET
2023	1 CHIC4	EXET CCT PHASE U/B	OFF
2023	AXMI4	X120 /	CLOSED
2023	MANN	1280/	OPEN

REFERENCES

1 Hein, F., "Expert Systems using pattern recognition by Real Time Signals", CIGRE, Paper 39.10, 1986.

2 Koike, N. et al, "A Real-Time Expert System for Power System Fault Analysis", IASTED International Conference on High Technology in the Power Industry, 1986.

3 Talukdar, S.N. et al, "TOAST The Power System Operators Assistant", I.E.E.E. Computer magazine, Vol. 19, No. 7, July 1986.

4 Sakaguchi, T. et al, "Prospects of Expert Systems in Power System Operation", 9th Power System Computation Conference, Cascais, Portugal, September 1987.

Chapter 16

Intelligent Process Control

Marylene De Winter, Marcel Rijckaert
Chemical Engineering Department, KU Leuven
de Croylaan 2, 3030 Heverlee, Belgium.

16.1 INTRODUCTION

Expert systems are computer programs which specialize in answering questions and solving problems concerning a particular domain. The questions asked and problems to be solved are of such a degree of difficulty that they require expert knowledge.

An expert system typically consists of three parts: a knowledge base, an inference engine or control structure and a human-machine interface. The knowledge base is the part where the experts' knowledge is stored. This knowledge can be expressed in rules or a similar expandable knowledge representation, saying which actions to take under certain conditions. The inference engine or control structure is the method that activates the rules. It builds up the reasoning process independent of the contents of the knowledge base. The human-machine interface is the module that makes it possible for the user to communicate with the expert system. The dialogue between the user and the expert system is held in a semi-natural language style. This makes it possible for the domain-expert to work with the expert system without going through the difficult stage of getting familiar with a programming language.

The most crucial difference between an expert system and a conventional computer program probably lies in the fact that the knowledge is inter-changeable and completely separated from the inference engine. Not only can you change the data your program is working with, but also the knowledge, i.e. the program itself, without affecting the inference engine. You can add or delete knowledge (rules) to your system without considerable effort if, for some reason, that appears to be necessary. Another difference lies in the way solutions to a problem are found. The solutions are not found in an algorithmic method, but by applying rules which are likely to lead to a satisfying answer. The solution to that problem may be a set of possible solutions. The expert system can jump to conclusions based on rules of thumb, just like human experts do. This means that even if the solution found is not the optimal one, it is an acceptable or most likely solution. Conventional computer programs can either provide a correct solution to a problem or no solution at all, they do not provide in-between, most likely solutions.

Furthermore most expert systems have an explanation facility that is able to explain "how" the solution to a given problem was found, i.e. which knowledge was used to come to the solution, given a set of data. In some cases the expert system will ask the user for additional data needed to apply a specific rule. At this point the user can ask the expert system "why" it is asking for this additional data.

Expert systems can be used in divergent fields of expertise. Most expert systems fall under one of the following categories:

- Interpretation which includes interpreting signals, analysing images and understanding speech.
- Diagnosis going from medical diagnostics to fault diagnosis in a chemical plant.
- Prediction systems, which infer likely consequences from given disturbances.
- Design, configuring objects, like computers or houses, given a certain set of constraints.
- Planning and scheduling or what you could call the design of actions, examples here are automatic programming and strategic military planning.

Still other applications use a combination of aspects from these different categories. Process control involves interpretation of signals, diagnosis of behaviour, prediction of consequences, planning of actions to be taken etc. By process control we mean all the actions that have to be considered to observe a process and to let it work within prescribed boundaries and limitations.

All these different application domains have very different requirements in hardware, software and manpower. Among the most demanding areas of applications is the area of real-time applications, such as process control. With real-time application we mean that the system should guarantee response to external stimuli in a given (usually short) time. Why are these applications so demanding?

First there's the speed of execution. We want our system to respond as quickly as possible, because most disturbances require quick action in order to prevent a catastrophe. Second there is the huge amount of data which must be recorded and updated very frequently. Third there is the difficulty of establishing a reasoning process that deals with time-dependent information.

Despite the fact that building an expert system for process control is not such an easy task, more and more people have become involved in it. In this chapter we try to point out the advantages an expert system can bring to the control of the plant in general and to its operator in particular. We also talk about the different tasks an expert system for process control should be able to perform.

This chapter focuses on process control for a chemical plant, which is the area of our research. Nevertheless, all the ideas are valuable for any type of process control.

16.2 WHY AN EXPERT SYSTEM FOR PROCESS CONTROL?

(a) Be up to date

In a world where speed and quality are sought at the lowest cost possible to fulfil economical requirements, the need was felt to create something new that would help to control the plant faster, better and cheaper. These were the original motivations for developing expert systems for process control.

(b) Assisting the operator

According to Zimolong et al., the tasks an operator has to perform can be divided into three categories.

- First, there are the routine tasks. The operator is very familiar with them because they happen frequently. The action to be taken is overt and does not require special knowledge from the operator: experience is all that is needed. (an example would be the support and checking of measuring-instruments).
- Second are the more complex situation tasks where the operator has to perform a set of routine tasks according to some time schedule or conditions. Usually the conditions to be satisfied and the time-schedule to follow are pre-planned by someone else. This means the operator has to be very well aware of the system's state in order to perform the actions correctly.
- Finally there are the unfamiliar and unstructured situations. They emerge when no routines or pre-planned structures for solving the problem are available. These are problems that are too complex or which, for some reason, require special treatment. In these cases what is needed is general knowledge about the behaviour of the system, characteristics of the environment and definitions of the goals we want to obtain. Problems arise because the operator has to gather this knowledge while surrounded by flashing lights, ringing bells, printers ejecting pages and pages of parameter values. There is so much data that it is almost impossible for the operator to figure out which parameters require immediate, undivided attention and which process parameters are related to each other. This overwhelming amount of information during an alarm situation makes it very difficult for the operator to perform his task without confusion. An expert system could give assistance here by quickly providing the operator with the information needed about the system. It could advise the operator of the actions to be taken next in order to solve the problem as quickly as possible limiting the damage caused by the fault.

(c) Controlling and supervising complex systems

A chemical plant is a very complex system. A lot of parameters have to be taken into account, where the relation between these parameters is not always obvious or linear. Complex systems are usually controlled by conventional programs. These programs are usually based upon linear models of the process parameters, or upon mathematical approaches of deviations of these parameters. The first problem that arises is the fact that some things cannot be captured in a mathematical model. Some units are just not measurable or computable. Second, if some mathematical model exists it is either too complex, or too simple. One has too make a choice between a multi-dimensional model that takes into account all of the parameters and possible deviations but is too complex to work with, or a simplified model that is easy to work with but does not include cures for the serious disturbances. Both systems will probably fail when a serious event, like the tripping of a boiler, occurs.

An expert system, whose knowledge base is the representation of the plant, could be a solution to the problem. This knowledge would be a combination of the operator's experience and analysed historical data. These are usually easier to represent by rules than by mathematical models. The conventional program can be kept for small, frequent disturbances, for the management of the individual units and the relation between these units based on non-dynamic behaviour. The expert system can be activated in case of an unexpected or infrequent event. It can supervise the dynamic behaviour of the system.

(d) Turning the knowledge of operators into a capital asset

The operators of a chemical plant are experts in solving the plant-problems. They have gained their expertise through years of experience in dealing with plant-problems. They are most likely the only persons who have sufficient knowledge about the plant to solve almost any problem. They are vital to the organization of the plant. What happens to the plant when the expert is no longer present? Even if he/she has attempted to communicate his/her knowledge to the successor, the expertise that made the operator so valuable is no longer available. An expert system could prevent this loss of crucial expert knowledge. Once stored in the knowledge base of the expert system, the expertise of the operator is available forever. If the knowledge base is regularly updated with information from newly encountered situations, after a few years it will turn into an irreplaceable gem.

(e) Training of new operators

Another advantage of having an expert system is that an elaborated expert system, containing expertise and solutions to problems set up by operators during many years, is a perfect tool for training new operators. One could set up a few simulations of trips and ask new operators which action they would take under the given circumstances. The new operators could then consult the expert system to have an overview of actions that have been taken under similar conditions in the past.

16.3 WHAT SHOULD AN EXPERT SYSTEM FOR PROCESS CONTROL CONSIST OF?

For an expert system to be effective for the control of a chemical plant, it must be able to perform a few different tasks. First and most important is the ability to handle serious disturbances. Second, it should have the capability to note trends and to predict disturbances in order to take actions that will avoid or reduce the damage that may be caused by the disturbance. Finally it should check for safety and economical efficiency of the actual configuration of the plant. These last two tasks are very specific for process control. We feel that they should be a part of any expert system that deals with the control of a plant. However, these tasks are very difficult to integrate into an expert system because they require very specific knowledge from the operators, knowledge they may never have formulated before.

(a) Handling serious disturbances

The first thing the expert system should be able to do is assist the operator in handling serious disturbances like the tripping of a boiler. When such an event occurs the operator should have a complete view of the plant in order to see what consequences this disturbance will have on the other units of the plant. The expert system should give this overview of the plant to the operator and determine the most appropriate response. It must quickly look at the states of the different units, check for a loss of steam-flow, determine which other boilers can be driven up in order to meet this loss, determine which valves have to be opened or closed to avoid an explosion etc. The expert system will then provide the operator with these infos and advise an appropriate action to be taken. This makes the situation less stressing for the operator and will reduce the errors caused by stress or ignorance of the total situation.

We recognize here one of the advantages of an expert system pointed out earlier in this text. The information and advice provided by the expert system is represented in a way closely related to the representation the operator has of the plant. It is *qualitative* information. Note that this is very different from *quantitative* information which is usually provided by conventional programs based on mathematical models and with which the operator is not so familiar.

(b) Predicting alarm situations

The expert system should be able to recognize and identify potential alarms.

How can this be done? The expert system should keep a history of the values that it is reading at specific time intervals. This means that if the pressure of the steam in a certain boiler has risen considerably without any obvious reason during the last ten measurements, the expert system must inform the operator about this. It should warn him that, if this rise of pressure continues, the boiler will probably trip. It can also give advice to the operator of the actions to be taken in order to limit the damages this trip might bring, or maybe even to prevent it altogether. Another possibility is to let the expert system find out the cause of the rise. As the knowledge base of the expert system is a simulation of the plant, it should not be so difficult to find what causes these sudden disturbances.

It is clear that all these occurrences must be handled with the greatest care. Before alarming the operator about a predicted disturbance, the expert system must have verified that it indeed concerns a real alarm and not a false one. It is possible that a pressure-curve of a boiler goes up for a few seconds, but automatically falls back again a few seconds later. It would be nonsense to alarm the operator in such situations.

(c) Checking for efficiency and safety

It seems apparent that the first two tasks described above enhance the safety of the plant. They reduce the risk of damage to plant and personnel by automatically shutting down part(s) or the entire plant when process variables move out of their normal operating ranges. They provide the operator with crucial information, thus reducing the possibility of making mistakes.

How can the expert system be used to optimize safety and efficiency of the plant control? An expert system (like the one we are building in Leuven) could, in periods where no disturbances occur and no dangerous trends are noted, simulate trips or other disturbances in order to see what effect this would have, considering the actual layout of the plant. For example, suppose the plant runs in a specific configuration and nothing goes wrong. The expert system could at this point build up a simulation of a boiler-trip and analyse the consequences this would have. This analysis may show that the actual design is not a very *safe* one, because e.g. the tripping of a boiler may result in a serious explosion. It can also analyse *efficiency* of the actual design of the plant. For example the tripping of a boiler may lead to a loss of steam in the amount that cannot be covered by the other active units. This means the layout of the plant is not an efficient one. The expert system could then advise the operator of the possible change(s) to the design, which would, in case of a trip, not result into danger of an explosion or irrecoverable loss of steam. This way the expert system provides the operator with up-to-date information which will enable him to implement optimum control strategies.

It is important that the modules which take care of predicting alarm situations and which check for efficiency and safety are only activated when no alarm situation is present. If the expert system is busy noting trends to see if anything is likely to go wrong and suddenly a boiler-trip occurs (someone pushed the wrong button) the handling of this trip should have absolute priority. The prediction-model should be interrupted and the alarm-module should be activated.

The same is required for simulation of trips. If, while a simulation is running, a serious problem occurs, the simulation must be interrupted immediately and the appropriate module must be activated. The handling of disturbances should always have the highest priority, then the notation of trends and finally the checking for safety and efficiency.

16.4 INTEGRATING THE EXPERT SYSTEM IN THE CONTROL SYSTEM OF THE PLANT

It should be clear by now that the primary motivation for installing an expert system is to enhance, and not replace, the traditional control software of the plant and the capabilities of the operator.

How could the expert system be integrated in the existing control structure? Typically a traditional system would measure the values of parameters of the different plant-units and transmit these values to the control software. There they are compared to set-points and small process deviations are automatically corrected. The process computer communicates with the control room so that the operator can have detailed overview of the real-time status of the process. Figure 16.1 illustrates how an expert system could be integrated in such a traditional control system. The data collector generates all the data which might be needed by the expert system. When the reasoning process of the expert system comes up with a conclusion, this is communicated to the expert system operator display, but in such a way that only relevant data is shown in order not to confuse the operator. The operator can then ask for a justification of this conclusion if that is necessary.

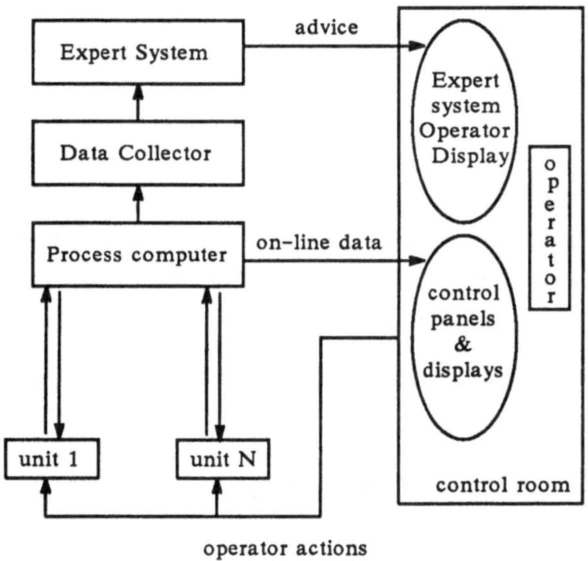

Figure 16.1. Integration of the expert system in the control structure of the plant.

16.5 CONCLUSIONS

An expert system for process control could be a very helpful tool for the operator of a plant. The operator is the person who has to provide the knowledge for this expert system. Although this is not the subject of this chapter it should be stressed that the elicitation of this knowledge is one of the most important tasks in the development of an expert system. The expert system can only be a valuable tool if the knowledge embedded in it is valuable.

The advice the expert system gives to the operator should be very selected. It is unnecessary and completely superfluous to overwhelm the operator with yet more flushing lights, ringing bells and beautiful curves. What the operator needs is clear and structured advice which makes to actions to be undertaken obvious.

Finally we would like to point out that an expert system for process control can be economically very effective. If the expert system prevents as much as one trip in the first year of installation, it pays back for its development price. The next catastrophes it prevents are pure benefit.

REFERENCES

1 Boullart, L., Delbar, P., "Expert driven industrial process control" , Proceedings of the forum on Artificial Intelligence and Chemical Process Control, organized by the Branche Belge de la societe de chimie industrielle, November 1987.

2 Hayes-Roth, F., Waterman, D.A. and Lenat, D.B., "Building expert systems", Addison-Wesley Publishing Company Inc., 1983.

3 Rijckaert, M., "Expert Systemen en proces controle", Technology Transfer Express, Benelux, No. 50, December 1987.

4 Rijckaert, M. and Bogaerts, W., "The incorporation of knowledge-based aids within the control and operation of a powerplant", December 1987.

5 Schindler, M., "Artificial intelligence begin,s to pay off with expert systems for engineering", Electronic Design, August 1984.

6 Shirley, R.S. and Fortin, D.A., "Developing an expert system for process fault detection and analysis", InTech, Vol. 33, No. 4, April 1986.

7 Zimolong, B., Shimon, Y.N., Ray E.E. and Gavriel, S., "On the limits of expert systems and engineering models in process control", Behaviour and Information Technology, Vol. 6, No. 1, 1987.

Chapter 17

NEW TECHNOLOGY FOR IMPROVED QUALITY CONTROL AND SECURITY OF PROCESS OPERATIONS

D. J. Sandoz
Vuman Ltd

17.1 INTRODUCTION

A modern control room will incorporate computer systems for the monitoring and control of the process, ranging from powerful large machines to simple PLC units. Process operation is typically a blend of automatic and manual supervision. The computers carry out startup/shutdown sequencing, a level of automatic control via conventional PID regulation and a level of supervisory control via purpose calculations. Color screen/keyboard facilities provide operators with information to manually oversee plant status and also the capability to interact with the control systems to ensure operation is to specification. Two important factors in an operator's brief are:

- To maintain product quality, at appropriate throughput and with appropriate economy and
- To ensure that plant operation is maintained within defined boundaries.

In the main, computer systems currently only provide aids to these objectives and it is necessary for operators to exercise skill and maintain continuous vigilance in order to ensure satisfactory continuity in operations. This human factor can give rise to considerable variability in plant performance and to occasional crises that lead to off specification production or worse.

This short chapter is concerned with the control engineering aspects that are employed by the computers and considers how these aspects may be progressed to simplify the supervisory role that is required of an operator and to improve the integrity of operation in the face of upsets from the plant environment.

17.2 CONVENTIONAL REGULATION AND ITS LIMITATIONS

The three term PID controller is a well known and long established control unit. Despite dramatic evolution in implementation, from pneumatic instrumentation to computers, the feedback technology for the majority of the industrial automatic control systems has hardly altered. It is simple and effective in many situations. Certain applications may involve hundreds of such controllers.

The PID controller has the straight-forward form[1]

$$u = P \quad E + \quad \frac{E}{I} dt + \frac{DdE}{dt}$$

with u the actuation, E the error, P the Proportional Gain, I the Integral Action Time and D the Derivative Action Time.

There are variations on this form, for example to avoid differentiation of error, but these are not normally of a major character. Most plant instrumentation personnel have an understanding of this algorithm and a certain knowledge and experience of how to set the parameters PID for appropriate controlled performance. This aspect is a strong factor in explaining the long term resilience of the three term controller.

The PID controller is ideal for controlling aspects such as flow, pressure and temperature. These are normally characterized by very simple first order lag dynamics. PID control is effective in controlling the supply environment to a process – for example feed of raw materials, steam flow/pressure control, preheating temperature control etc. It is however, much less effective in controlling the product characteristics of a process – for example moisture, density, quality indices etc.

Many industrial processes involve the transport of material through successive processing stages. A large evaporator, for example, may have as many as six or more successive evaporation units. It may typically take as many as 15 minutes for material to propagate through the evaporator. A PID controller may be used to control product density in such a situation. However, for stable operation, the proportional gain would have to be very small (because of the transport delay). Thus, although the controller would hold steady state under stable operating conditions, it would be ineffective in recovering that steady state subsequent to some major disturbance to the plant. It is in such circumstances that manual intervention would be essential in order to protect and maintain process operations. Difficulties of this nature have often meant the abandonment of automatic control, with product control effected manually by the process operator. In any event, such product control has to be manual in many circumstances because the properties of the product that require regulation cannot be measured continuously. Operators maintain control on the basis of regular laboratory analysis of product.

There has recently been rapid progress in instrumentation technology so that the online measurement of product quality is now feasible in many circumstances. Since, in many cases, PID control cannot be effective in controlling product quality, an alternative and more progressive approach to control engineering is necessary if the level of plant automation is to advance to more generally cover product quality.

17.3 TECHNIQUES THAT CAN ADDRESS PRODUCT CONTROL

There are two significant technologies that are evolving that can address the product control situation:

- Rule Based Control, that employs expert system concepts so that the computer is able to reflect the behaviour of the skilled operator and hence replace him [2]

  and

- Predictive Control, that employs models of the process established by application of engineering science and statistics [3, 4].

Experience in the application of Rule Based Control is relatively slight. The Cement Industry has been the main area of experimentation with certain successes claimed. The approach involves compiling a data base of plant operation, in effect a catalogue of IF/THEN/ELSE rules that characterize all situations and define actions to be taken. Rule interpretation may involve techniques such as Fuzzy Logic to extract decisions from situations where a variety of choices may be open. At the moment the technology seems best suited for the steady-state supervisory control situation rather than for dynamic feedback control. For example, Rule Based Control does not yet address time delays in a satisfactory manner (other than on the "nudge and wait" principle). However, it is an exciting technology that touches on the arena of Artificial Intelligence. It will evolve strongly as more potent techniques for intelligent interpretation of rules evolve.

Predictive control is a more mature technology that has evolved from the systems engineering developments of the 1960s and 1970s. It directly addresses plant dynamics and offers significant advantages over conventional regulation. The most simple form of Predictive Control is the Smith Predictor. This is a well known approach for time delay compensation that involves the process transfer function with a conventional controller. However, such control is not widely used, probably because of the very specific engineering skills that are required if application is to be successful. The more general concept of predictive control offers a more attractive facility for advanced control.

Predictive control relies upon a model that is an approximate representation of the cause and effect relationships that describe the plant that is to be controlled. Establishing such a model is the first phase in design. Least squares statistical techniques, in application to data sampled from the process, provide a basis for determining the coefficients of an appropriate model. Computer packages are now available to provide the engineer with these statistical tools for process modelling. The modelling exercise is interactive. The engineer selects simple parameters, such as time delays, sampling intervals etc., and steers to a satisfactory result by making successive trials and evaluations.

Given a model of suitable accuracy, analysis then progresses to develop the predictive control system itself. There are a number of methods for such control system design. The most effective, in the experience of the author, is that which involves the minimization of a quadratic function of error. The analytical techniques to achieve such minimization are complex. However, such complexity is an issue of software rather than one of engineering use. Again, computer packages are available that implement this design technology. The user need not be aware of the analytical detail but may steer the analysis, via an interactive approach, to achieve control system performance that is satisfactory. Such performance may be appreciated in simulation by applying the designed controller to the model.

Given a predictive controller derived by the above approach, this may be implemented in real time in application to the actual plant. A difficulty is that the engineer is not able to relate directly to the coefficients of such a controller. Controller gains have no physical interpretation, in contrast to the conventional control system gains P, I and D. It is therefore essential that additional facilities are available in support of a predictive controller so that an engineer may retune the controller should performance deteriorate. Such facilities require a real time implementation of the modelling and control system design techniques outlined above, and in fact comprise the essence of a self tuning or adaptive control scheme.

Thus, the successful application of predictive control requires the process engineer to be presented with considerable design and real time software. If the process engineer is not to be an expert in the design technology (essential if the technology is to be widely used) such software must be very user friendly and must disguise the complexities of the analytical techniques employed.

The benefits of a predictive controller are:

- The controller may predict through time delay. Thus tight control is not compromised by the presence of time delay.
- The controller may compensate for interaction between plant areas, eliminating hunting problems (i.e. multivariable control).
- The controller may predict and eliminate influences from plant disturbances (i.e. feedforward control).
- The model may act as a noise filter to permit steady controlled operation in the face of adverse signal to noise ratios.

The above properties make the predictive controller a much more effective vehicle for the automatic quality control of product than the conventional PID control.

Studies in the application of predictive control, for example with moisture control in spray drying and temperature control in cement kilns, have indicated that substantial cost benefits may be achieved because of steadier operations and improved throughputs. Steadier control in particular is a major feature in the safety of plant operation. Crises induced by the variability of manual control are avoided.

17.4 A SIMPLE COMPARISON OF CONVENTIONAL AND PREDICTIVE CONTROL

To emphasize the clear advantages of Predictive Control, a simple comparison of capability is presented. Figure 17.1 illustrates the stimulated step response of the product density (signal M20) of a multiple effect evaporator to a step change in the feed flow rate to the evaporator (signal S2). A change of 40 kilograms/hour infeed rate gives rise to a change of approximately 30 kilograms/cubic metre in density. The response covers some 15 minutes and has the character of second order dynamics. Figure 17.2 illustrates the implementation of predictive control to minimize the impact of feed flow changes upon density. The steam flow rate to the evaporator (signal S5) is varied to maintain the density at a defined setpoint. This variation is based upon model prediction and employs both feedforward and feedback action. Note that following a change in feed rate the steam adjusts promptly and prior to any change in density. Density variation from setpoint following a disturbance is about 8 kilograms/cubic metre.

In contrast, Figure 17.3 presents the setup response of a well tuned PID controller to a step change in density set point. The response has a well defined character with some 5% overshoot and an acceptable transfer time. On the face of it, the PID control system appears quite satisfactory for plant application. However, if the plant is subjected to the same disturbance conditions as indicated in Figure 17.2, a different picture emerges. Figure 17.4 illustrates the performance of the PID controller in such circumstances. The density now swings wildly from setpoint to more than twice that of Figure 17.2 (around 20 kilograms/cubic metre) and the time for recovery is greatly extended.

In reality, density control on multiple effect evaporators is almost always a manual affair because of the inadequate closed loop capability typified by Figure 17.4.

17.5 DISCUSSION

The comparison of Section 17.4 clearly highlights the disadvantages of conventional systems for product control of industrial plant. Product quality measures (e.g. density, moisture, chemical composition etc.) are very commonly seen to be under manual control. This human factor in plant control gives rise to constraints in performance and weakens the security of operation. Plant efficiency and product quality vary from operator to operator and also, for a particular operator, with the attention employed to the control task. It is also possible, and a frequent occurrence, for an operator to inadvertently engineer a crisis in operation because of poor plant management. This can lead to plant shut down if safety boundaries are exceeded.

There are therefore clear economic and security justifications for automatic product control. Conventional technology cannot usually address product control, as indicated above, however, the newer technologies of predictive Control and perhaps the Expert System approach of Rule Based Control do offer a basis for a more progressed level of automation and plant security than is currently achieved.

REFERENCES

1 Warwick, K. and Rees, D., "Industrial Digital Control Systems", Peter Peregrinus, 1986.

2 Haspel, D. and Taunton, C., "Application of Rule-Based Control in the Cement Industry", Blue Circle plc and Sira Ltd.

3 Clarke, D. W. and Gawthrop, P. J., Automation Vol. 17, No.1, 1981.

4 Sandoz, D. J., "CAD for the Design and Evaluaton of Industrial Control Systems", Proc. IEE Vol. 131, No. 4, 1984.

5 VUMAN CONNOISSEUR, Documentation, Vuman Ltd, Enterprise House, Lloyd Street North, Manchester.

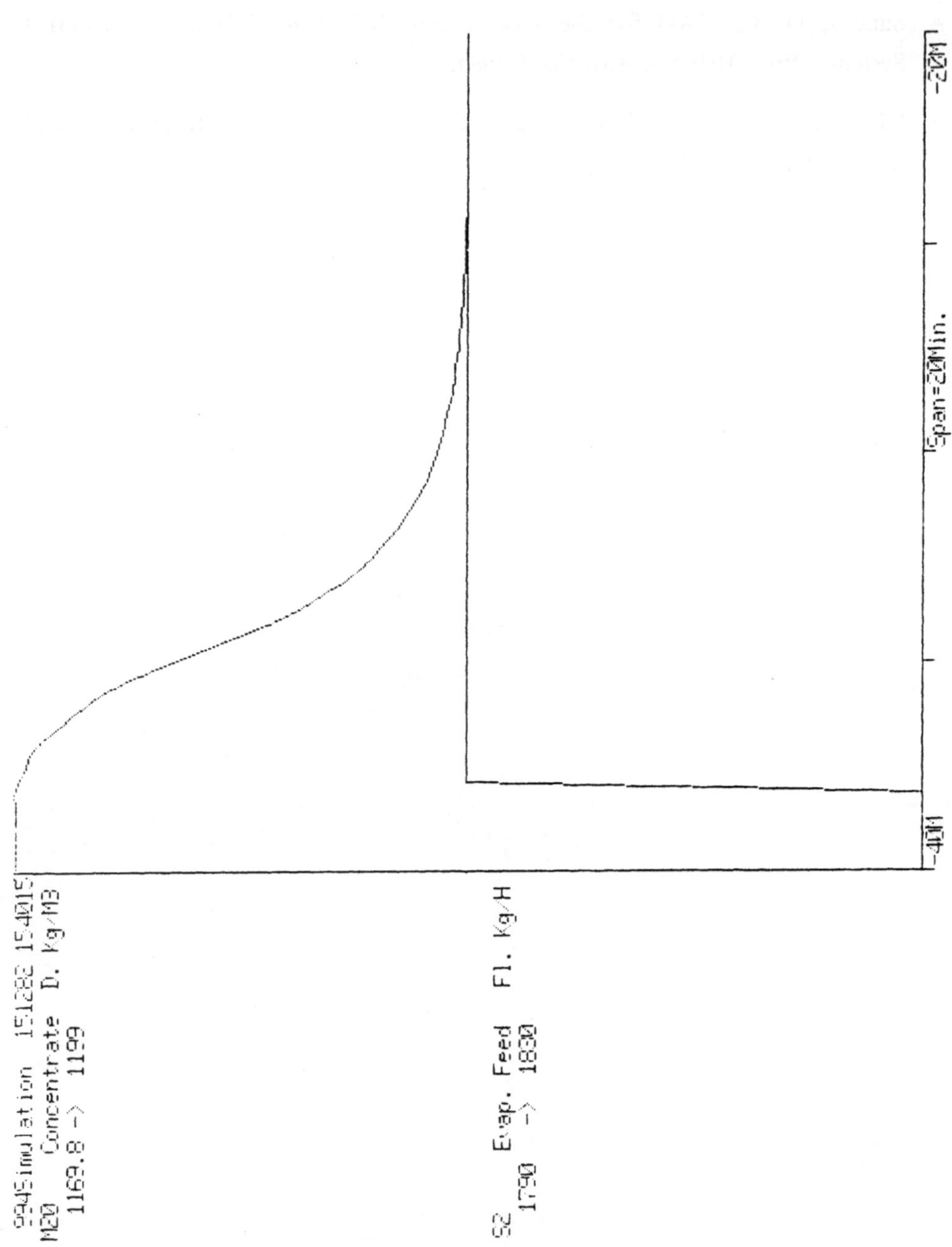

Figure 17.1. Response of density to a step change in feed flow.

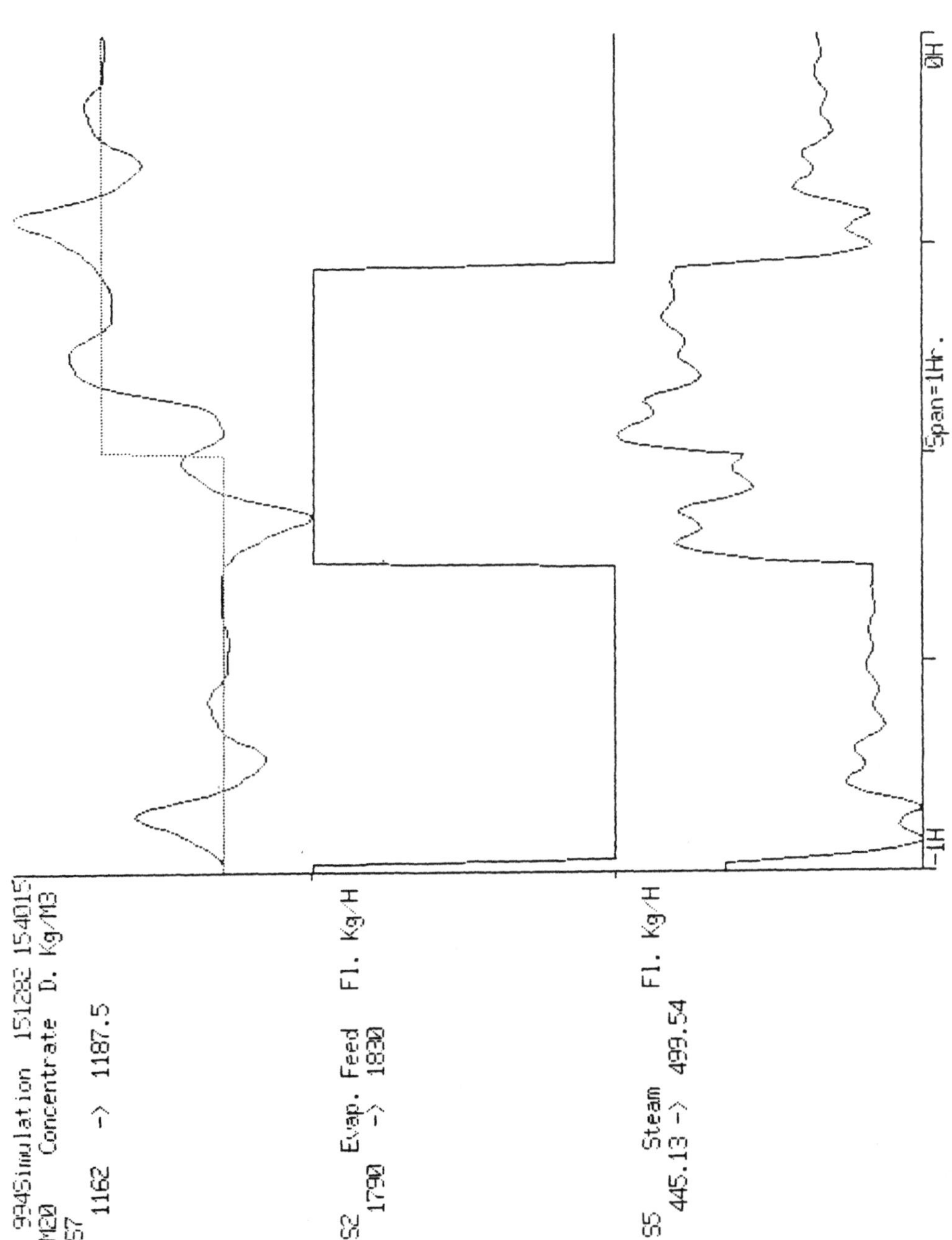

Figure 17.2. Predictive control in action to minimize impact of feed changes.

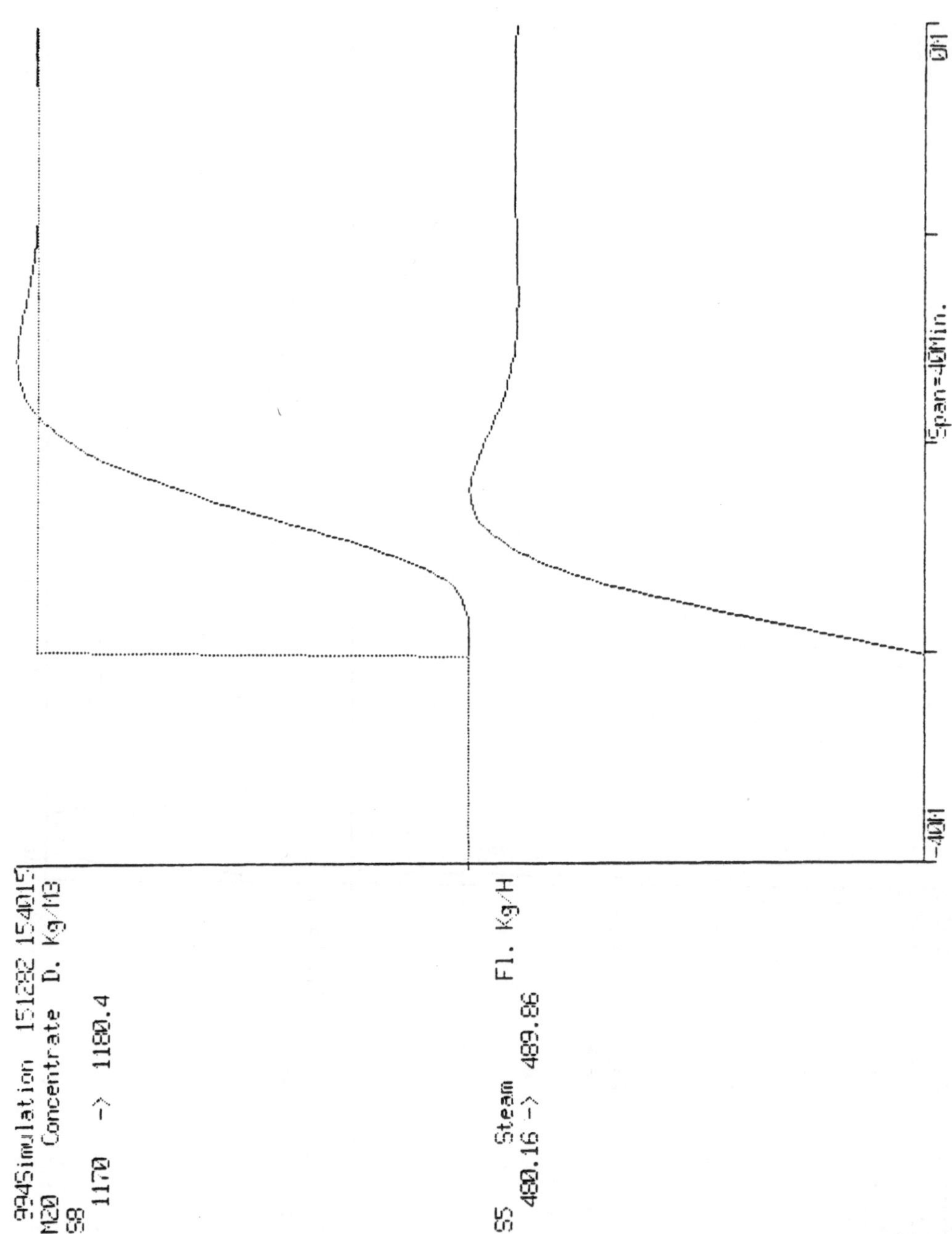

Figure 17.3. Response of a PID controller to a change in density setpoint.

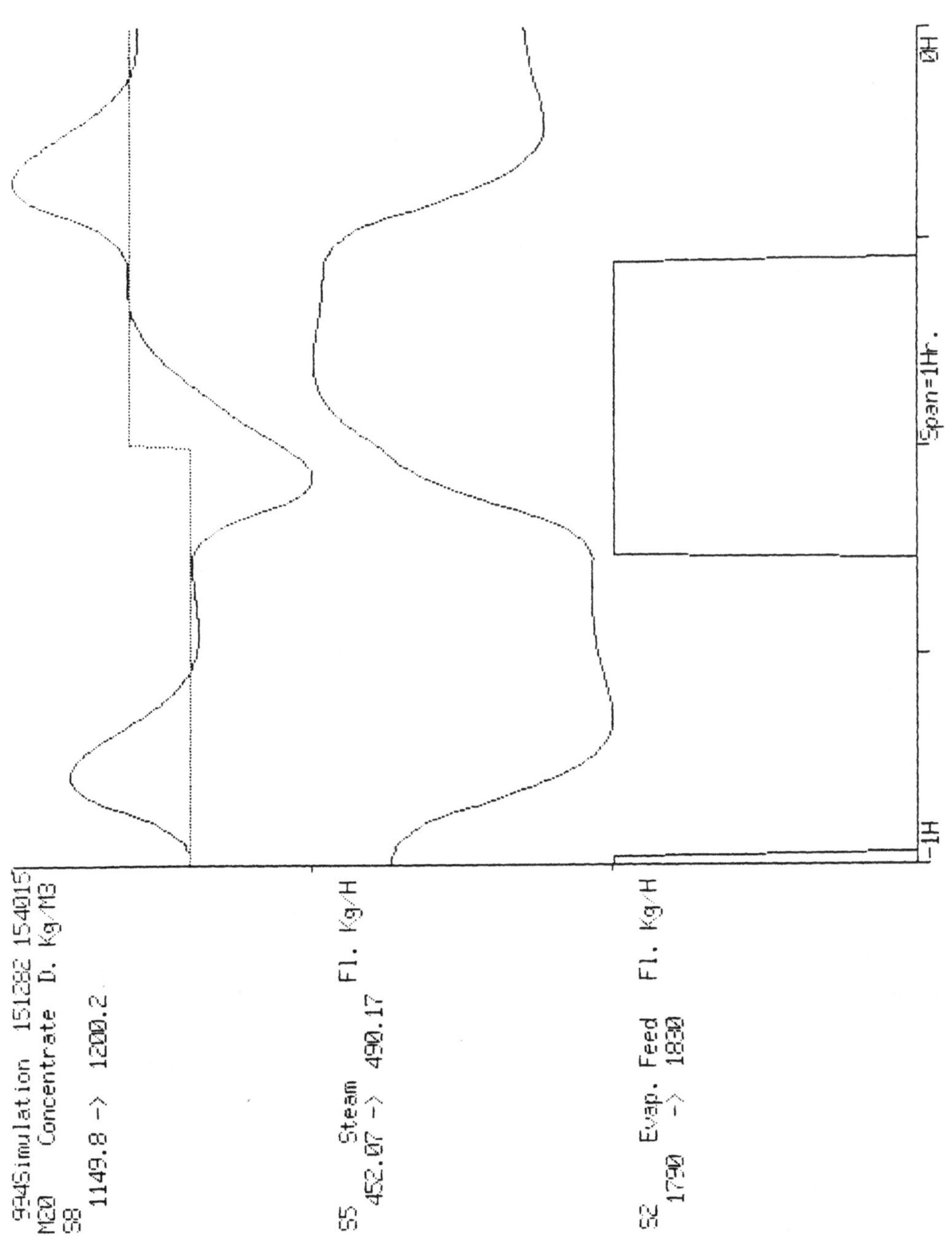

Figure 17.4. Performance of a PID controller in the face of feed flow changes.

INDEX